GEOWRITING

a guide to writing editing, and printing in earth science

second edition   edited by

Wendell Cochran
Peter Fenner
Mary Hill

American Geological Institute
Falls Church, Va.   1974

credits

The concept of this manual originated with one of us [Fenner] as a result of evaluating a report stressing need for better writing by geologists. As that report, *Requirements in the field of geology* (by David M. Delo and Robert G. Reeves; 1969), was published by the American Geological Institute's Council on Education in the Geological Sciences, the matter was taken up with the Council's Professional Development Panel, then consisting of Reeves (chairman), Donald R. Baker, Donn S. Gorsline, Keith F. Oles, and Ralph B. Ross. That panel approved a project that would concern not only writing but also its extensions—editing and printing.

Much of the early work was done by members of the Association of Earth Science Editors, notably Gerald M. Friedman, Richard V. Dietrich, Patricia Wood Dickerson, John W. Koenig, Arthur A. Meyerhoff, Thomas F. Rafter Jr., William D. Rose, and Martin Russell. At a late stage in the writing, additional advice was provided by an AESE group consisting of Albert N. Bove, George E. Becraft, and A. F. Spilhaus Jr.

As this manual evolved, material by contributors (most of them members of AESE) was divided up and redistributed, so that specific credit is difficult to give. However, we thank Robert L. Bates, Richard V. Dietrich, John W. Koenig, and Robert Mc Afee Jr. (for critical reading), Trinda L. Bedrossian (symbols), Elizabeth Bennett and Mary Beth Cumming (typing and proofreading), Jules Braunstein (style and usage), Kenneth L. Coe (abstracts), George V. Cohee (geologic names), Patricia Wood Dickerson (references and usage), Edwin B. Eckel (whose views on refereeing survive virtually intact), Peter Fenner (for his chart of geologic time), Josephine F. Fogelberg (copyediting), William H. Freeman (his section on book reviewing remains almost as he wrote it), Douglas M. Kinney (geologic maps), Kenneth K. Landes (whose classic work on abstracts is reprinted by permission from the *Bulletin* of the American Association of Petroleum Geologists, v. 50, September 1966, p. 1,992), Joel J. Lloyd (abstracting and indexing), Melba W. Murray (who wrote the sections on organizing reports and on personal style; the first is a modification of her paper "Written communication—a substitute for good dialog", published with permission of the Esso Production Research Company, Houston, in the *Bulletin* of the American Association of Petroleum Geologists, v. 52, November 1968, p. 2,092; the second is adapted from that paper and also from her book *Engineered report writing*, revised edition, 1969, Petroleum Publishing Company, Tulsa; these sections are published here by permission of Esso Production Research Company, Petroleum Publishing Company, and the American Association of Petroleum Geologists), Mark Pangborn (maps and bibliographic matters), Willis A. Shell (printing technology), Donald G. Turner (typography), Richard J. Vorwerk and Joseph C. Meredith (future methods), B. H. Weil (for excerpts from "Standards for writing abstracts", reprinted by permission from *Journal of the American Society for Information Science*, v. 21, 1970, p. 351; copyright 1970 the American Society for Information Science, Washington, D.C.), and Sherman A. Wengerd (manuscript preparation).

Staff members of CEGS were helpful throughout, especially F.D. (Bud) Holland Jr., and William H. Matthews III (successive AGI directors of education), and Jackson E. Lewis. All along, the backing of the CEGS Council was a necessary adjunct; during this time the Council consisted of George R. Rapp (chairman), John E. Bowen, Peter Fenner, Jacob Freedman, Donald Groff, Warren D. Huff, Raymond Pestrong, Robert G. Reeves, Ralph B. Ross, Joseph L. Weitz, and Donald H. Zenger.

Special thanks go to the National Science Foundation for its financial support in preparation of the manuscript and for printing.

Despite all that very considerable help, responsibility for opinions, interpretations, and the like remains ours.

Wendell Cochran (Geotimes)
Peter Fenner (Governors State University)
Mary Hill (California Division of Mines & Geology)
August 1973

In the second edition we have corrected a few errors pointed out by readers, added a new section ('getting started'), and rearranged the order of some sections to make a more logical sequence.
March 1974

Geowriting: a guide to writing, editing, and printing in earth science, edited by Wendell Cochran, Peter Fenner, and Mary Hill. CEGS Programs Publication 17. Copyright © 1973, American Geological Institute, Falls Church, Va.

International Standard Book Number 0 913312 02 9.
International Standard Serial Number 0084 9375.
Library of Congress catalog card number 73 85048.

This book was photocomposed in Photon Optima (the body type is 9 on 11) by Bru-El Graphic Inc. of Springfield, Va., and printed (on 60-lb Beckett Brilliant Opaque) and bound (the cover is 65-lb Navajo) by William Byrd Press of Richmond, Va., in the United States of America. Design and production by Wendell Cochran.

year 987654
edition 98765432
printing 987654321
3M TOLL

contents

start here

This book is an introduction to writing, editing, and printing, especially as applied in earth science. It should help you understand a process that starts with writing a manuscript and ends only when the work is printed, bound, and reviewed. It is a how-to-do-it manual only in part, for every topic discussed here has been dealt with exhaustively elsewhere. It is instead an outline and a guide—a resource intended to give a notion of problems, tools, methods, references for digging deeper. It should be useful to:

• students (undergraduate or graduate) approaching for the first time the possibility of writing for publication

• professors and supervisors who foresee that their students or employees will someday write or edit

• geoscientists who publish infrequently and so find it difficult to keep in mind all the related mechanical matters

• scientists who find themselves—sometimes unexpectedly—appointed or elected to editorships.

We intend this as a functional work. Our style decisions are those that seemed best suited to this manual alone (given the biases of the editors) and are not necessarily meant as standards for other works. We hope that the examples and discussions will help other writers and editors make the choices most suitable for their own publications.

In general, each section of this book stands alone—it does not depend on earlier sections. The arrangement follows, roughly, the mechanical order of publication: writing through editing to printing. Writing, editing, and printing are not really independent entities; each can exist without the others, but if any is treated that way some potential is lost. To a craftsman, writing-editing-printing is an interlocking whole.

preparing typescript

- *Match article and journal.*
- *Follow that journal's style rules.*
- *Type with uniform spacing.*
- *Keep a copy for yourself.*
- *Double-check all names, spellings, arithmetic, references . . .*
- *Use a simple, meaningful title.*

Most editors are harried, some are lazy, and none likes to do needless work. As a writer you should take advantage of those traits: make things easier for the editor and you'll see your work in print sooner than otherwise.

Market your article. Find which journal is most likely to use it; don't submit a paper unless you are familiar with recent issues of the journal and are really sure your paper will fit it better than any other. For example, the *Journal of paleontology* and the Geological Society of America's *Bulletin* may both use a paper on Devonian brachiopods, but the *Bulletin* is less likely to use one on a specialized aspect such as the morphology of brachiopods.

Editorial policies change, so make certain that you have examined a *recent* issue of the journal; that is, do not depend on an old issue or on your accumulated impression of the journal's content.

If the journal you choose has a style book or other standard, get a copy and follow its instructions closely. Some of the rules may seem merely whimsical, but follow them anyway—nearly all are designed to fit an editorial system or to meet mechanical requirements imposed by that journal's design or the printer's equipment. Flouting those rules makes work for the editor and tends to prejudice him against your paper.

If the journal has no formal suggestions to authors, apply the rule of common sense. Use *some* style book. Keep everything simple and consistent. Avoid such typists' devices as single-spacing the abstract, long quotations, and references in an attempt to simulate smaller type.

Although many style books specify that manuscripts (purists prefer "typescript") must be submitted on "good quality white bond paper" (some go so far as to specify the weight), that appears to be mere superstition. Editors must work with the paper, and they object to paper too stiff to handle, too thin for convenience, too shiny for easy reading, or too hard or slick to take sharp, clear pencil marks. Papers with easy-erase surfaces are particularly hard to work with. Test the paper. Can you write on it easily without smearing the writing or tearing the paper? Can you use an eraser? Will the paper fade or discolor in ordinary use?

rules for typist

All copy must be typed—double-spaced—with normal indentations, wide margins (at least 4 centimeters) and approximately equal line lengths. That means *all* copy, with no exceptions; it includes titles, by-lines, author identification and affiliation, abstracts, quoted matter, footnotes, tables, lists of references—everything. Double-space it all, and do not add extra spaces between lines, paragraphs, or sections. (Triple spacing is acceptable—but be sure to use the same spacing throughout. Single spacing is never acceptable.) Avoid block paragraphing—that is, with no paragraph indentation for the first line.

Do not break a word at the end of the line, as the editor or typesetter may mistake the hyphen for a part of the spelling.

Use one side of the paper only. At the top of each page, type a *short* identification tag—one word, such as your last name—and the page number, as Jones 1, Jones 2, Jones 3. It will help the editor if at the bottom of each page except the last, you type "more"; at the end of the last page, type "end" or "30".

Impress those admonitions on your typist. The article will be easier to type that way, and the editor's work will be easier: the typist will not waste time changing settings on the typewriter or calculating indentations, and the editor can use the space between lines and in the margins to correct spellings, give instructions to the typesetter, and the like. Also, equal spacing and even margins help the editor estimate how much space your work will occupy in the publication.

Try not to use a proportional-space typewriter, which makes copy-fitting—estimating space—all but impossible. (For that reason, some editors return any articles typed on such a machine.)

For most editors, the original and one copy will suffice, but be sure to keep a copy for yourself. Note that some copying machines use paper that is hard to write on; test all paper with a soft black pencil.

Photographs and other artwork must be identified in such a way that they will not be damaged; perhaps the best way is to write the information on a separate piece of paper and *tape* it to the back of the artwork. (Be sure to include location, scale, and any pertinent credit.)

If you can incorporate footnote material in the text, do so.

estimating space Some editors welcome an outline of a prospective manuscript; it can aid both writer and editor because they can then discuss manuscript length in terms of the number of words, or of typescript pages, or of printed pages. As a guide: one printed page in this book has about 550 words, or about two pages of double-spaced typescript.

Avoid built-up fractions, such as this one:

$$\frac{a-b}{3(a+c)bx}$$

for they are expensive to set in type and waste space. Where possible, use case fractions, like this: (a–b)/3(a+c)bx. (That form is easier to type, too.)

Most science journals now use the metric system, and you can speed the editorial process by following suit.

Double-check all spellings, quotations, references, equations, formulas, and arithmetic. It is the editor's duty to check such things, too, but he can't always find the paper you quote from, nor can he check "personal communication" and "in press".

Use a person's name the way he commonly uses it himself if at all possible. For example, reducing "M. King Hubbert" to "M.K. Hubbert" complicates the literature and invites typographic errors ("M.L. Hubbert" is more likely to get past the proofreader than is "M. Ling Hubbert").

In tables, avoid rules—horizontal or vertical lines. Remember that the editor can add needed rules far more quickly and neatly than he can take unneeded ones out. Use enough space between columns to make the meaning clear but not so much as to make lines of numbers hard to follow with the eye.

Do not abbreviate names or terms, especially journal names (unless your journal specifically requires it). Leave that to the editor; it's his job, and he should not have to find whether your "Geol." means "Geologic", "Geological", or "Geologist".

Do not attempt to specify design or to mark typographic style except to italicize species names and the like. That, too, is the editor's job.

permissions If you plan to quote from published material, or to reproduce another author's photographs, tables, or diagrams, be sure to get permission in writing. (Many publishers have standard permission forms, which simplify and speed this task.) Obtaining permission is a matter of legal obligation to the

- *Leave design and layout to production people, and let editors do the editing.*
- *Use some style guide.*
- *Consider production stages and the relative costs of changes at each stage.*
- *Keep a current copy of whatever you send to the editor; include corrections that have been made or requested.*
- *Select a short descriptive title and write an abstract (both must be suitable for computer indexing).*
- *If you are concerned with copyfitting, consider the specifications before you get too far along with writing.*

copyright owner and also a matter of courtesy. At the very least, a conscious effort to find where permission may be required will help avoid the embarrassment of accidental plagiarism.

If you need to acknowledge help, be brief and straightforward. Fulsomeness and the passive voice waste time and space, so be direct: write "Joseph White helped me greatly with the field work" and not "Extreme gratitude is due to Joseph White for his valuable aid and advice in the field." (Why not give him the gratitude instead of continuing to owe it?)

Watch for trademark names, such as Plexiglas, Geodimeter, and Xerox. Carelessness in spelling and punctuation can result in sharp letters, and worse, from lawyers.

Writing often differs from speech, so if you have an article for publication and also plan an oral presentation of it, consider revising it so that you can give it comfortably aloud.

Editors sometimes must modify titles of papers for typographic and other reasons. You can help both editor and reader by keeping the title of your article brief and specific. Remember the key-word requirements of computer-based bibliographies, and avoid such words as "introduction", "principles", "selected", "investigations", "Recent", and "recent".

Finally, if you want your work back, say so, and enclose an envelope with your name and address and sufficient postage.

This book contains many check lists intended to ease the path from concept to publication. If you use them—or construct your own to fit your needs—your worries will be lessened. Make your first check list as soon as you start writing, or even before.

getting started

"When you can't paint, paint", advised artist Gulley Jimson in *The horse's mouth*. Similarly: when you can't write, write. Never mind the grammar, the syntax; the logic and organization, the one word with the exact shade of meaning, but concentrate on getting down on paper everything you know that is pertinent to your aim in writing. Then take a pencil and edit your work. Using scissors and tape, rearrange paragraphs or whole sections. Edit it with a pencil and run it through the typewriter again. Repeat as needed.

The procedure just described is emergency treatment, to be used only for a severe case of writer's block. If at all possible you should outline before writing, expand the outline, write, and then edit, rewrite, and polish.

Editing in this sense involves a process somewhat like that of using a zoom lens, in which you zoom down to a tiny detail, back up for the big picture, and zoom down again. When you change a word or phrase, take a moment to consider the effect on the whole sentence, then on the paragraph, and then the entire section or even the complete report. (If you change present tense to past, ask yourself: What else here is affected?) When you transpose paragraphs, check the transition and watch for antecedents.

The analogy applies to proofreading, too. If the typesetter has corrected an error, make sure he corrected it properly; also make sure that he did not introduce another error elsewhere in the same line, that the corrected line is in the right place, that the defective line really was removed, and that no adjoining lines were disturbed. Only then will it be safe to zoom back down on the next correction.

Even this zooming is not enough. As a writer you must repeatedly look at your work as the reader will see it, anticipate the reader's questions, and take steps to prevent them. Here an old newspaper rule is useful: never underestimate a reader's intelligence; never overestimate his information. You are an expert in the thing you are writing about; your reader may not be. You have reference works before you as you write; probably your reader will not. You know exactly what your sentences are supposed to mean; your reader may be puzzled. So, when editing for rewriting, take the reader's point of view.

You have at least one other role to play: that of editor. He should be your most critical reader, anticipating the questions of the ultimate reader and prodding you for more information here, clarification there. He will also have constantly in mind the problems of production—the myriad details of transforming your words into the final form of ink on paper. Even though he will probably know much more about those things than you do, you can save much time and trouble by trying now and again to see things as the editor sees them.

All that is to stress the point that the boundaries between writer, editor, and printer are illusory, and that interchanging roles will help everyone achieve the common goal: serving the reader.

putting it together

Good organization
- *helps tell a message clearly*
- *is satisfying to reader and writer*
- *may be obtained through various approaches.*

Literary demands on today's professional scientist are not what they were 75 years ago, when relatively few people were sufficiently well educated to impress readers with polysyllabic words and complex grammatical constructions. Present-day interest is in ideas, not words; the essential message, or "news", of a report or paper must not be buried in stuffy prose. The emphasis is on ability to communicate efficiently, which means clearly, briefly, and economically. This emphasis is plainly evident in industrial recruiting practices, efficiency rating and promotion policies, and cost-control procedures.

To present a tidy set of rules for meeting these literary requirements is virtually impossible. No two writing projects present the same set of writing problems so far as organization of complex ideas is concerned, and no single standard outline or format can be offered to make organization easy. However, it is possible to suggest guides that can be adapted to suit the writer's problems, habits, and capabilities, and that will ensure a logical presentation of ideas, hold down length, and reduce frustration.

The conversational approach to organization is a set of guidelines, in the form of questions based on the pattern of good dialog, with which you can decide logically what goes where in a paper, article, report, or letter. The method itself is unconventional, but the documents produced by the method range from the "traditional" to the "new"; they vary infinitely, as do the people reading and writing them. Whatever the outline, whether different or familiar, the message can be almost as direct, as natural, and as easy as face-to-face dialog.

The method is based on some simple facts. To communicate is natural and easy. We all do it with some skill. And we demonstrate it daily as we talk with people. In conversation, where we are aware of people as live bodies and responsive minds, where we are conscious of practicalities in the realm of ideas and concepts, we are as clear, as brief, as simple, as direct, and as forceful as the occasion demands. We organize ideas instinctively in response to human needs. Writing is simply a substitute for this more desirable means of communication.

We can depart, then, from the traditional practice of writing suspensefully, using as an invariable standard the pattern that proceeds from Introduction through Results to Conclusions and Recommendations. We can simply use the *pattern of good dialog* as a *disciplinary guide* for prose that is based on good judgment rather than rules. Briefly, the idea is to pose the questions a reader might ask in conversation about your subject (the silent half of the written dialog); provide the answers you would give in conversation (the oral half); use these answers to construct an outline that gets directly to the point in every major unit; use an expansion of this outline to dictate or write a draft in the unpretentious language of conversation; and finally, edit and polish. The key to this conversational approach is advance planning, which begins with the people involved in the "dialog".

begin with readers Analyze these people—your readers. This analysis is a critical preliminary to actual writing, one we always do in theory but rarely in practice. *Making this analysis in writing* is a self-imposed discipline the value of which cannot be imagined until it is tried. A simple questionnaire may be helpful in this respect.

Who will read it?

The needs, interests, and background of the readers will tell you how to organize. Do not generalize in your answer; give names of people. It is easy to communicate with specific people. Lacking a real audience of human beings, we write to ourselves as a substitute, and for this kind of writing there is little demand. If you are not sure of names, find out. Ask questions. If you have a broad public audience, such as the members of a professional society, identify two or more persons by name. If you are writing for a professor or other students, write to them specifically.

What is their interest?

Why would they read what you write? How can they make practical use of your information? Does it solve a problem for them? Can you save them time or money? Can you give them information or an analysis or synthesis they would otherwise have difficulty obtaining? Have you improved methods or procedures they can use? Have you created a service or a product they need? Have you expanded a body of scientific knowledge with information that can be directly tied to their professional interests or needs? (If none of these, why write?)

What is your interest?

reasons for writing
What is your purpose in writing (or your company's, or your professor's, in asking you to write on a subject)? Do you expect readers to act or to buy a service, product, or idea? Should readers simply read, understand, and respect what you are adding to their knowledge? In any case, what approach would induce a reader to do, think, or feel the way you want him to? Knowing who he is and what he wants, you can decide how to approach him before outlining, and thus eliminate many outline possibilities. For instance, you know that a general reader is usually repelled by tedious experimental details, but a researcher in your field might be moved to action by such details.

Do you have problems? How will you handle them?

A most urgent job in writing is to settle inhibiting problems of tact, diplomacy, and policy, before outlining. Tensions induced by hampering security regulations, by fear of skepticism or antagonistic reception, by uncertainties in the data, or by any of many factors that tend to "freeze" the writer, must be relieved before composition. Only the author who faces these problems honestly and decides how to write so as to handle them is free to concentrate on his prose and able to converse in simple, natural style.

developing the message
Use the pattern of dialog to develop the message. Another questionnaire might be used as a discipline for getting to the point—the essential "news", not only for the report as a whole (the opening summary or abstract), but for each major unit of the report as well. This discipline reduces length 25 to 50 per cent without losing desirable content. Consider how you might ordinarily outline an article (for instance, Abstract, Introduction, Literature, and Experimental Procedures). Then imagine how you might inform a reader if you were talking to him in the hall. The leading questions might go something like this: "About that problem . . ."

What's the news?

Why? (Why should I care? Why did you do that? Or why should I do that? How does this tie in with my problem?)

How? (How do you know that? Or how did you do it? Or how should I proceed to do that? Explain it. Support it. Illustrate it. Make it believable. But don't give me all the details, please; I'll ask for your working papers if I need them.)

Now what? (What, then, is my present position—or yours—or ours? What next?)

These questions do not constitute a standard format. Not all of them would be asked in every conversation, nor would they always be asked in this order. Still, it is probably safe to say that most of the time most readers would like to get to the news at once. In dialog it would rarely seem logical to hold up the message in order to relate details concerning it. Details will naturally decrease, in any case, if taken out of the introductory position and used instead as support. Only the author who knows his audience and his message can anticipate the nature or relative urgency of a reader's inquiries, and he will vary the order of his questions (vary his organization) accordingly.

build on the outline   Construct an expanded outline. Using the questions and answers that represent the conversation, first a topic outline and then an expanded topic outline can be constructed. Both will maintain the pattern of conversation, and assure that the message takes precedence over details.

● Jot down in the margin, beside each answer, the topics that the answers suggest. Then number them to show rank, weight, and interrelationships in the usual way.

● Under each major topic, insert a summarizing topic sentence that answers the question "What's the news?" about that unit. Here is an example of a summarizing topic sentence: "The most important variables to consider in defining this function are porosity and depth; others, such as temperature and elevation, are unimportant." The next example is only a topic sentence and gives no "news": "Several variables were considered to determine which would be most important in defining this function."

The outlines below show how this method worked for a seismic interpreter writing a report on an areal survey he had made for an oil company. Here is his first effort:

suspense outline

    I.   Introduction
        A. Past exploration history
        B. Geologic data available from well logs
        C. Magnetic maps
        D. Gravity maps and residual anomalies
   II.   Method of analysis of seismic data
        A. Seismic field methods
            1. Processing of seismic data
            2. Filters
            3. Types of display
        B. Cycle correlations
  III.   Stratigraphy
        A. Stratigraphic column
            1. Preliminary stratigraphic units
            2. Probable lithologic types
            3. Possible stratigraphic traps
        B. Description of cross-sections
  IV.   Tectonic history
        A. Types of structures
            1. Salt tectonic areas
            2. Down-to-basin slump faults
            3. Possible structural traps
        B. Description of cross-sections

V. Results
   A. Types of structural prospects
      1. Down-to-basin slump faults
      2. Salt-dome crest anticlines
      3. Other fault-trap types
   B. Types of stratigraphic trap prospects
      1. Reefs
      2. Updip pinchouts on margins of salt domes
      3. Updip pinchouts on unconformities
      4. Fractured shale
   C. Operational problems in area
      1. Near-surface-layer seismic anomalies
      2. Terrain
VI. Summary, conclusions, and recommendations
   A. Geology of area
   B. Location of drillable prospects
   C. Recommendations
Appendices
   A. New solution to gravity anomaly calculations
   B. Maps and seismic cross-sections with overlays

In this easy but dramatic situation the "news" that was buried in sections V and VI of this outline can be expressed quite simply: "It's a bonanza! Let's start drilling!"

topic outline
    Here, in the more formal language of technical writing, is the dialog and the topic outline based on it:

The silent question	The written answer	The conversational outline
What?	I recommend that the area outlined in blue on Map X be acquired immediately for intensive exploration.	I. Recommenda-tions
Why?	As Map X shows, this area has an excellent potential for oil accumulation. The two best prospects are a major salt dome and a large reef with 2,500 feet of closure. Nine other prospects include other reefs, salt domes, gravity slide faults, up-to-the-basin faults, and fractured limestone in an intensive fault zone. Reservoir facies include turbidite sand, beach or alluvial sand, reef limestone, and shallow-water, shelf-type limestone.	II. Summary of prospects

How?	These prospects are revealed by data from three E-logs, one dip log, two seismic sections, and a gravity and magnetic map of the area.	III. Supporting data
		  A. Salt tectonic
		  B. Reefs
		  C. Faults
		  D. Stratigraphic traps

Now what?	In view of the enormous potential of this area, we should lease the property at once and start drilling the salt dome and reef. At the same time, we should start additional shooting to detail the other prospects.	IV. Recommended program

expansion
And here is the topic outline with summarizing topic sentences, or an expanded outline:

I. Recommendation
   I recommend that the area outlined in blue on Map X be acquired immediately for intensive exploration.

II. Summary of prospects
As Map X shows, this area has excellent potential for oil accumulation. The two best prospects are a major salt dome and a large reef with 2,500 feet of closure. Nine other prospects include other reefs, salt domes, gravity slide faults, up-to-the-basin faults, and fractured limestone in an intensive fault zone. Reservoir facies include turbidite sand, beach or alluvial sand, reef limestone, and shallow-water, shelf-type limestone.

III. Supporting data
These prospects are revealed by data from three E-logs, one dip log, two seismic sections, and a gravity and magnetic map of the area.

A. Salt tectonics (Figure 1)
1. A salt intrusive, perhaps domal in outline, has created traps from the oldest beds to the youngest.
2. There is a salt withdrawal pillar with traps in the older beds.
3. A salt overthrust anticline with traps is developed in the other beds.

B. Reefs (Figure 2)
The flanks of the reef area in the older beds contain traps created by pinchout and by closure over the top.

C. Faults (Figure 3)
Large tectonic faults involving the older beds offer trap possibilities. In addition, down-to-basin slump faults create closure on the downdip sides.

D. Stratigraphic traps (Figure 4)
In addition to stratigraphic traps associated with structure, a belt of sand pinchout is indicated across the upper area of the map.

IV. Recommended program
In view of the enormous potential of this area, we should lease the property at once and start drilling the salt dome and reef. At the same time, we should start additional shooting to detail the other prospects.

Bear in mind when writing college papers, journal articles, research reports, and sometimes industrial reports as well, that the expanded outline may need to provide headings or subheadings for documentation and proof of replicability of results. In writing, as in conversation, such data fit most naturally after the claim or statement they support. If you examine the seismic interpreter's expanded outline above, you will note that there is a natural, conversational place for such data in every unit of the How section.

After completing the report, if you find yourself with a stack of leftover data you might consider disposing of it (a) by inserting a short unit where it least intrudes, (b) by moving the data to an appendix, or (c) by omitting it as "unnecessary detail". The (c) option has shortened and strengthened many documents without depriving their readers at all.

confer before writing   Call for a pre-report conference. If your paper or report must be released by supervisors, management, or professors, send each one a copy of the expanded outline and ask for a conference a few days later. A good expanded outline may be completely adequate to get the technical content and proposed format of your report approved before it is written.

Dictate and revise. At each unit of the expanded outline, begin with the opening summarizing topic sentence and then dictate or write a rough draft of whatever is left to be said. With the "news" already expressed in the opening sentence, it will be quite clear how much—or, more important, how little—will be needed in the way of support, development, or explanation. The ready-made lead sentences also make the job of dictating much easier for beginners in dictation.

edit and polish   In editing, you should check your verbs, provide any missing transitional signals, scan summarizing topic sentences to be sure they make your points at the opening part of each unit. You should be very sure that the reader can clearly distinguish among facts (yours and others'), suppositions, inferences, and conclusions. If all your data cannot be given with equal certainty, modify and explain to account for the differences.

The conversational approach to organization does not set a standard outline, nor does it eliminate the various kinds of outlines to which you may be accustomed. The chief idea is to avoid allowing any single standard format to strangle style and function—delivery of a useful message in your own best voice. The idea is to consider every written message individually, applying good judgment and good taste rather than rules with respect to its user, its subject, and its purpose.

The outline given in the foregoing discussion is only one of many possible models. Other models are equally valid. For instance, a skillful analysis of the reader (or a publication) may properly lead to a more traditional organization with headings such as these: abstract, introduction, literature review, methods used, results obtained, discussion, conclusions, recommendations, acknowledgements, bibliography. Some journals specify such an outline. One can still apply the What, Why, How . . . questions at each major unit of the outline, keeping the reader's informational needs in mind, and reduce the length of a paper dramatically.

end with readers   Regardless of the organizational scheme thus developed, know and relate to your audience, remember that the same story should be told differently to different people, and get to the point.

four kinds of style

"Style" can mean many things:
- *editorial or style-book style*
- *literary or personal style*
- *typographic style*
- *word usage.*
No style fits every publication.

When a writer or editor refers to "style" he may mean any one of many different things. This is a matter of practical semantics, because fuzzy boundaries make for fuzzy application and inefficient communication.

Briefly, "style" usually means one of four things:

1. Editorial (or style-book) style concerns primarily matters of consistency in such things as capitalization, abbreviation, punctuation, use of numerals, and the order of components within bibliographic citations.

2. Literary style (or, as E.B. White put it in *The elements of style*, "the sound that words make on paper") relates to euphony, rhythm, avoidance of obtrusive alliteration and accidental rhymes, and the like. It also concerns personal preferences and habits of expression, and so is sometimes called "personal style".

3. Typographic style involves the typefaces, sizes, measure, and so on, as used in any particular publication.

4. Style in the sense of usage is used loosely to mean style-book style or literary style, or both. More strictly defined, it deals with definitions, connotations, idiom, and such distinctions as those between "eager" and "anxious".

Most publications and organizations have their own editorial style rules; many have their own style books. These commonly contain guidelines that apply to all the classes listed above and thus contribute to the confusion. As one result, style books are notoriously hard to use.

The difficulty of organizing, indexing, and using style books has led the compilers of some of the larger ones to adopt a strict alphabetic order; in the style books of the *National Geographic magazine* or the *New York Times*, for example, you do not look under "dates" but under "November" to find whether the preferred form is "November 1" or "Nov. 1".

There can be no such thing as a perfect style book; as Edward Sapir, a noted linguist, once said, all grammars leak. All styles, too.

Although style books are full of whims, such as a preference for "grey" rather than "gray", the basis of any style must be found in the subject matter of the particular publication involved.

Whims aside, no very detailed style book could be devised for all the various fields commonly lumped under the name "geology". Paleontologists use the term "new species" so often that repeated spelling out with species names would be intolerable, but geochemists see the term so rarely that "n.sp." must not be abbreviated in their journals.

However, any writer should adopt (or adapt) *some* style for his own standard. Style books agree in the larger matters more often than they disagree, and so learning one style—almost any style—means some measure of general consistency. Further, learning one style makes it easier to note the differences and thus easier to learn another style.

finding a style book

If no particular style has been ordained for the writer or editor, he should first examine the style books (if any) in his general field. For example, a paleontologist might find that he could easily adapt to his needs the *CBE style manual* (by the Council of Biology Editors). As a news magazine, *Geotimes* uses an adaptation of the *New York Times style book*. In any case, the writer or editor should mark up his style book freely to change the rules as time goes on and the accumulating evidence shows what is the most efficient for his writers, editors, and readers.

Literary style, being less dependent than editorial style on the subject matter, has been covered extensively in works on writing. Anyone in doubt of where he stands at this point should start with chapter 5 of *The elements of style* and take it from there. Also, any specialist who deals in writing and editing should assemble his own collection of clichés, and examples of careless usage and inept syntax, taking care to eschew the merely arbitrary "rules" so common in such lists.

A less conventional approach to literary style may be found in *The art of readable writing* by Rudolph Flesch. In it Flesch attempts to *measure* the readability of prose; even if you do not agree that his rating system works you are sure to pick up tips for making your own writing easier to read. (A few extreme examples: Use short sentences. And properly chosen incomplete sentences. When you can, use personal words like "you". But don't go overboard.)

Usage is closely related to literary style and often overlaps it. Here, the great majority of problems are covered in such works as *Modern English usage, The careful writer,* and *The complete plain words.* Geologists need not repeat all the work that went into those books; they should devote their attention to the problems peculiar to their own field. Again, it is useful to compile a list from your own experience of inappropriate, imprecise, and inefficient usage—provided that you review the list periodically with an unprejudiced eye.

The rules for typographic style are usually fixed for a given periodical. A pure example is hard to find in published form, for they are seldom useful to anyone but the editors of the publications involved (but see pages 110–124 of *The New York Times style book*).

codifying style

If an editor new on the job finds that his typographic style has not been set down in writing, he should start right away to set it down if only as a convenience to his successor. In doing this, he should confer with his typesetter and printer (that will also greatly improve his communication with them and add to his understanding of his own publication). Also, a written typographic style is essential when the time comes to obtain comparative bids for typesetting and printing.

A common but primitive approach is to leave it up to the printer to "follow style" based on previous issues. That course virtually bars improvements, for when the editor tries to change a style point the printer is likely to consider the change an error. Worse, the practice encourages the printer to believe that the editor does not know his own business. He is likely to be right.

For a new or one-time publication, style conventions often seesaw. An editor might decide that his journal should consistently use the form "per cent", but then after publishing a few papers with statistics requiring frequent use of the term change his style to the form %. Later he might find that statistical papers had become exceptions and that "per cent" (or perhaps "percent") was more fitting after all.

That occurred in the course of work on this book: each contributor naturally used the style of his own publication, but as material accumulated, the editors found that some of those rules fit the total content and that others did not. Also, there was a certain amount of give and take, one editor perhaps arguing for a general policy of Latin forms rather than the Anglicized version (symposia versus symposiums), another for the shorter form of a word often spelled two ways (dialogue versus dialog).

As an aid to determining style, we editors made up a short list of style questions as we went along—a list that others may find useful:

- Capitalization or lower case—National Geographic Society vs. National Geographic society; the Editor or the editor.
- Italics or quotation marks—"Modern English Usage" or *Modern English usage*, for titles of books, periodicals, and chapters.
- Quotation marks—'English' or "American".
- Commas, periods, and quotation marks—it is common practice to place commas inside quotation marks, as in "The New York Times," in all cases, but some prefer to place them according to meaning—"The New York Times", for example.
- Abbreviations—Sacramento, California, or Sacramento, Calif. (or Cal., Ca., or CA); USA or U.S.A.; No. 1, no. 1, n. 1; 3 kg or 3 kg.; per cent, percent, %.
- Figures—eight, nine, 10, or 8, 9, 10; 14 6-point slugs or 14 six-point slugs; 1000 or 1,000; 28,000,000 or 28 million.
- Hyphens and diphthongs—over all, over-all, overall; and cooperate, co-operate, coöperate.
- Latin or English—curricula, curriculums; formulae, fomulas; symposia, symposiums.
- English units or metric.
- Plural forms—1970's, 1970s.
- Long form or short—employee or employe, gauge or gage, catalogue or catalog, dialogue or dialog.
- Other variant spellings, such as gray or grey.
- One word or two—artwork or art work; stylebook or style book; halftone or half tone; checklist or check list.
- Citation style—word order, abbreviations.

a style of your own

The best literary style is
- *your own*
- *concise*
- *clear*
- *simple*
- *active.*

speech as a guide

Style in writing and speech is a blend of one's personality, temperament, and individual peculiarities. A characteristic of good writing is that it delivers readers from monotony, dullness, and boredom. The writer should cherish his individual style and strive actively to preserve it. But along with this privilege to protect style is an obligation to the reader—to perfect it.

"Personal style" is not a good reason for poor communication. Unnecessary length and complexity, for example, should not be imposed on readers in order to indulge a style that needs improving. Readers are entitled to clarity and brevity—to writing that neither strains their patience nor overtaxes their time. They have the privilege of rejecting long, stuffy, or hard-to-understand writing, through procrastination if nothing else, and they often exercise this privilege at a loss to themselves as well as to the writer. Profitable ideas frequently go unread and unexploited.

Clarity and brevity can be incorporated into any style by sensible practices that, like logical organization, will alter style only to the extent that it improves the product.

Here is some high-flown rhetoric for you to consider: "Inferences may be extended to a reasonable position from which a decision may be made as to the advisability of collecting additional information." Meaning: you may need to collect more data.

Or: "The mineralogical techniques used in these studies were those which were determined to be the most advantageous from a standpoint of both speed and accuracy." Meaning: the mineralogic techniques were the fastest and most accurate available.

Or ask three geologists to interpret the meaning of this one and compare results: "These salt motive [sic] or salt accentuated anticlines may be responses to linear salt thicks and thins due to partial solution of salt in the X interval or effects of low relief linear topographic expressions present on the pre-salt rocks being slightly rejuvenated by differential warping or flexuring due to gulfward tilt."

Much technical writing is, like those examples, dull, stiff, or hard to understand—because of unnecessary complexity. What is meant by "unnecessary complexity" is perhaps best explained by what is *not* meant.

No reasonable critic quarrels with the use of technical language; without it a specialist could not even think clearly. A large vocabulary is essential, too; without the right words technical prose would be impoverished, imprecise.

Personal style has nothing to do with complexity. In fact, if every author insisted on using his own best human voice in writing, the technical literature would be enriched.

This is not to suggest that editors or employers invite exhibitionism. Slang and excessive levity would be inappropriate in a scientific journal. There are other limitations, too, such as the obligation to use standard English (constructions preferred by most educated people) and good dictionary terms that deliver the same message to all. Fortunately these rules are flexible and do not hamper style; "correct" usage is, as new grammars and the latest international dictionaries reflect, changing to meet changing demands on the language.

One can write (or dictate, if practiced in the art) with an uninhibited flow of words, then edit out the unnecessary complexity, and still retain a natural, individual style that keeps readers awake.

The chief thing to strive for is the simplicity that adds to the reader's understanding.

The same skill in writing that assures fluency sometimes invites excesses. For example:

the long version . . .

"In view of the necessity for the delineation of the contact between the X Formation and the Y Formation in the solution of the problem at hand, a program was initiated for sampling and studying the littoral sediments of the Z area under conditions which would assure adequate coverage of the interval of interest. The samples collected were subsequently subjected to detailed study. Although the conclusions reached may undergo alteration as more information becomes available, dependent upon the information obtained to date it is possible to establish an arbitrary base for the X Formation as the contact between an overlying series of interbedded red and greyish-green shales and fine-grained quartzitic sandstones and an underlying series of interbedded grey calcareous, exceedingly fossiliferous and grey non-calcareous, fissile shales, and thin-bedded, fine-grained calcareous sandstones, the latter series representing the top of the Y Formation."

Half a dozen working geologists first interpreted this text with some variations but finally agreed on a simplification:

. . . and the short

"To define a contact between the X and Y Formations, we sampled the nearshore sediments in the A area at 5-foot intervals. The sample study is not complete, but we can set tentative boundaries.

"The bottom of the X Formation is chosen as the base of a series of interbedded red and greenish-grey shale and fine-grained quartzitic sandstone. The top of the Y is a series of interbedded grey calcareous, very fossiliferous shale, grey noncalcareous fissile shale, and fine-grained calcareous sandstone in thin beds."

In the simplest analysis, complexity is related to sentence length and "big" (polysyllabic) words. The original passage and the revision compare as follows:

	number of words	average sentence length (words)	percentage of big words
original	146	49	26
revision	89	22	15
cut	40%	55%	42

A 40 per cent cut in length is significant not only to the reader, in terms of comprehensibility, but also to the publisher, in terms of cost. For instance, the average length of manuscripts submitted to the *Bulletin* of the American Association of Petroleum Geologists is about 9,500 words. According to the editors, a 40 per cent reduction in just one of these average articles would reduce the cost of editing, setting, and printing by more than $340.

brevity and readability

Note that the simplified and shortened version improves readability in other ways. It limits the number of ideas per sentence; admits personal pronouns where they fall naturally; and turns fuzzy expressions (adequate coverage) into concrete terms (5-foot intervals). There is a pleasant variation in sentence length, from short to long, but the average is low. Most important, the writing is unpretentious, with little words used for little ideas—conversationally.

active vs passive

Efficient use of verbs will do much for your writing and for your professional image. Verbs in the active voice are strong, direct, forceful, and economical, and they are natural to your conversational voice.

To illustrate, here is one of the sentences from the revision above, with an active verb (subject performing the action upon a grammatical object):

We sampled the nearshore *sediments . . .*

As the arrows show, the flow of action is forward. If we use the passive verb (which acts upon the subject of the sentence), the flow of action is backward, the sentence grows longer, and the pace is slower:

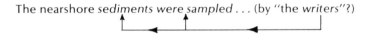

The nearshore *sediments were sampled . . .* (by "the *writers*"?)

The worst thing about passive verbs is that they often do not survive in this simple form; they are frequently distorted into other, unnecessary parts of speech. For instance, in the above sentence the verb *sample* would commonly be turned into a noun, like this:

"*Sampling* of the nearshore sediments . . . ."

and a new predicate would have to be found to complete the thought, like this:

". . . *was carried out* by the writers."

The original sentence has now grown progressively from 5 to 8 and finally to 11 words. Editors often reduce the length of reports by 10 to 50 per cent by revising with respect to verb usage only.

-ing, -ment, -ion

Whenever a writer finds himself searching for verbs like *was carried out, was done, was accomplished, was performed, was observed, existed, occurred, was present,* he should check to see if he has turned good verbs into other parts of speech, usually words ending in *-ing, -ment,* or *-tion.* For example:

"Water *flows* through the passage"; *Flowing* water [can be observed moving?] through the passages.

"Water *displaces* oil"; *Displacement* of oil by water [occurs?]."

"I *applied* for the job"; "*Application* for the job [was made by "the writer"?]."

The general rule in using active and passive verbs is to strive for a good, useful balance. The passive verb is both useful and necessary. It serves to show correct emphasis, to make smooth transition from one idea to another, and to provide variety.

dictating

One of the finest techniques for writing clearly and naturally (and for logical internal organization) is dictation. It gets into the prose the sound of the human voice (yours) and an original style (also yours, rather than a copy of what you think sounds "professional").

An author with no dictating experience should not expect good first results. Dictation takes practice. Some people never get the hang of it. It's a skill worth striving for, though, not only for its value as a clear-writing technique but also for its value as a cost-cutting device. One research organization has calculated that if all its 350 professionals learned to dictate reports only, the company would save over a million dollars worth of productive research time each year; in terms of manpower, the benefit would be an added 35 researchers without new hiring.

a final word . . . or two . . .

In summary and conclusion,

This chapter represents an organization and synthesis of material available to the author on techniques for achieving clarity and brevity without destroying style, and is presented for the utilization of readers interested in the expeditious preparation of good articles and reports. It is hoped that this compilation will not only provide a means of facilitating ease of preparation, but will also result in a considerable improvement in the quality of technical literature published in company reports and in professional journals. [80 words]

That is—

We hope these ideas will help you write good articles and reports more easily and clearly—in your own *best* style. [21 words]

drawings and photos

If you are preparing articles, your illustrations probably will be photographs, drawings, or maps. Those three classes overlap: there are paintings (oil, acrylic, pastel, watercolor) and drawings, in various media, that closely resemble photographs; there are shaded maps that are much like drawings; there are high-contrast photographs that resemble line drawings.

The division is not so much between the method of constructing the artwork as it is in the method of reproduction. Photographs are usually reproduced by the "halftone" method; color photographs require separation into the colors to be printed. Line drawings can be reproduced with the text material, as text, too, is a grouping of lines, all to be printed in the same color of ink. Maps, on the other hand, are of both types: some are to be printed in full color; some are simply line drawings.

Preparation of geologic maps is dealt with elsewhere in this book, so not much need be said about them here. Most geologic maps are printed from separate plates for each color, prepared from drawings or scribed negatives that the cartographer makes. However, it is possible to electronically scan hand-colored maps and to make color separations from the scan. It is also possible to make photographic copies of hand-colored copies of the map. One state geological survey did this recently in order to have full-size proof copies, in color, of a new state map. But the cost was nearly $100 per copy—expensive by anyone's standards.

Maps are sophisticated products that few people know how to make or use; but almost everyone makes photographs and many understand them. For that reason, perhaps, the photograph is the form of illustration most mistreated. Too many geologists are wont to point a shaky camera at a stunning subject, have the negative processed at a drugstore, make a small fuzzy print, draw on the front of the print with a ball-point pen, write a caption by hand in hard pencil on the back, attach it to the covering letter with a paper clip, and mail it to the editor in a plain, unstiffened envelope.

Good graphics are a must. Take the same care with the artwork that you did with the original piece of research. Good graphics and good research make possible a magnificent presentation.

A good picture is worth a thousand words; a bad one may say something different from what the author meant it to say. Geologists are lucky in that their subjects usually move slowly if at all. This may beguile them into thinking that no care is required to photograph them, but accuracy and clarity apply to each piece of artwork.

Most nonprofessional photographers stay too far away from their subjects. If you know each geologic unit thoroughly, communicate this intimacy in your photograph. Get close to the subject or plan to crop closely when you make the print. (But be sure it is clear which is the top.)

Wait for good natural light if you are taking exterior photographs, or add light if you cannot wait. A poorly lit photograph seldom shows the subject well, and can be only a poor illustration when printed.

Try to compose the photograph in an interesting manner. If you do not know the general requirements for good composition, consult a good "how to" art book. With care, anyone can take good photographs and thereby improve his scientific contribution.

Ask the editor of the journal you are writing for, or read the specifications of the journal, to decide what size of photographic print to submit. Generally, it is better to send photographs larger in size than they will be

Illustrations include
- *geologic maps*
- *black-and-white photos (halftones)*
- *line drawings*
- *color photos.*

Illustrations for talks should be specially prepared.

Illustrations for publication must be
- *clear*
- *neat and clean*
- *adequately captioned*
- *fully identified.*

taking the photo

when published. There are limits, though: a photograph that the printer must enlarge more than three times doubles the cost of the plate; if it must be reduced more than six times, that too can double the cost of the plate. Prints should generally be ferrotyped (given a high gloss) in order to display the maximum amount of detail.

handling photos   Cleanliness, neatness, and care are watchwords with all artwork. Once you have a good photographic print suitable for reproduction, treat it with great care. Do not fold, spindle, or mutilate! Don't dent it with paper clips. Never staple artwork. Identify it in some manner so that the editor can tell which photograph matches which caption, but refrain from writing on the back in such a way that the writing comes through as embossed lines on the front. To be safe, rubber-cement or tape a duplicate caption (with your name, date, illustration number, and article title) on the back of each photo.

Don't write on the face of the artwork, even if you would like lettering to appear on the printed picture. One reason is that your lettering would be screened or "halftoned" with the photograph.

If you intend to draw lines around formations, or to identify objects in the photograph, do the drawing or lettering on a transparent plastic overlay, clearly marked in such a way as to assure proper registration—that is, so that the marks on the transparent overlay will be printed exactly where you want them to be printed. Generally, little circles with crosses in them,⊕, are used for registration. (Art-supply stores stock them in adhesive rolls and in sheets.) Place them outside the margins of both pieces of art, one aligned exactly over the other. Three marks on each sheet should guarantee registration.

Art stores also stock letters and numerals in a variety of styles and sizes, as well as symbols and patterns, suitable for doing professional lettering on your overlay or drawing. Once you have made an overlay or drawing in this manner, treat it, too, with care. If the drawing has been done with anything that can smear, such as pencil or charcoal, spray it with fixative, and cover both the drawing and photograph with a protective sheet of acetate or paper. Remember that any writing you do, even erased pencil lines drawn for lettering, or words you have written on a sheet of paper on top of the photograph, may leave pressure marks that will show when printed, sometimes even if invisible to the unaided eye.

cropping   Be sure to obtain a clear photograph in good focus, cropped as you would like it. If the photofinisher has not cropped it enough, you can indicate further cropping to the editor by making a paper mask to go around the photo. This will not interfere with whatever system he uses for marking cropping limits and reductions for his printer.

Here are some suggestions for cropping:

1. Enhance the subject. Crop the photo to stress its purpose. If the caption reads "The rock is horizontally layered", then the best cropping, other things being equal, would be horizontal.

2. Crop tightly to the subject, maintaining good composition and photographic interest.

3. Crop to improve composition. Reduce distractive elements, focus on subject, improve balance and relationship of design elements in the photograph.

4. Crop to remove photographic blemishes.

5. Watch the scale. A graphic scale is by far the best; if one is indicated by an object in the photograph, don't crop out that object unless you indicate the scale in some other manner. If you must use a caption that gives a mathematical scale for the photo, be sure to maintain the scale in printing, or recalculate it to correlate with the new size.

6. Do not crop people out of a photo if they contribute to reader interest in the subject or if they indicate scale or depth of field. (But do not allow people to distract from the real purpose of the photo.)

7. Vary the shape of the photo if that will add interest to the page. Unless uniformity is a deliberate means of design, avoid it. Uniformity does not equal quality.

8. In general, make sure that "scenics" or photographs encompassing large areas will be printed larger than close-ups.

graphic scales   If you have a scale in the photograph, identify it carefully in the caption ("20-meter tree"). If you have no scale in the photograph, you may add a graphic scale to the overlay, or state the magnification or reduction in the caption. Your editor should take your mathematical scales into account when he calculates his printing enlargements and reductions, but he may forget or miscalculate—so graphic scales are greatly preferred.

What about retouching? If you can retouch well, you can probably improve your illustrations greatly. Use an airbrush to remove blemishes; dodge and burn when making photographic prints; use retouching pencils to clarify and improve detail. Nothing is taboo, short of changing scientific fact.

Line maps and figures—those that are not to be printed in color or shades of grey—should be drawn clearly enough to present a neat, legible appearance when printed. This usually means that they should be drawn larger than they will be when printed—perhaps twice as large. Be sure that any lines you make or letters you use are bold enough to remain clear after being reduced. If your budget will stand it, you can have your drawing reduced photographically or by electrostatic process to show you exactly how it will look when printed. Some editors prefer to use a reduced photographic negative and positive print rather than the original artwork. If your work is publishable without correction, this is an excellent solution; if not, you may be asked for originals.

If your budget cannot bear the cost of a negative, you can have your drawing reduced by an electrostatic copier to check your work. You may be sure that the printed illustration will be much better than the copy, but don't let faith in the printing process blind you to faults in your drawing. You may wish to submit only the electrostatic copy of your drawings (but never of the photographs) until the article has been accepted for publication. This will save wear and tear on your original artwork. However, the electrostatic copy will not be suitable for publication.

size and shape   There are several ways to determine what proportions your drawing or photograph will have when reduced. The easiest way is to use a reducing lens or a simple machine (you can buy either at art stores). Two other ways, described below, are graphic and mathematical methods. These methods will tell you whether your reduced drawing or photograph is the right shape, but not whether it will be legible.

Although the author should plan his drawings to fit the format of the journal he has selected, he should not mark reductions on the illustrations. This is the editor's job. The editor is responsible for marking the illustrations for reduction in a way that both he and his printer understand.

Some journals require a border around each illustration. Check with your editor to see if this is required; if it is not, decide whether your artwork will be improved by a border. (But too much bordering can kill an illustration as surely as if it were the lines around a funeral card.)

Many line drawings are small maps or cross-sections; turn to the check list for geologic maps to see how many of the items listed there will be needed on your map. How many should be on the drawing itself? How many should be given in the caption?

Here's how to figure reductions: to use the graphic method, make a dummy of the page shape as shown (reduced) in the margin.

Use a light-table. If the illustration is to be within margins, place the lower left corner of the illustration at "A", under the dummy. On the dummy, very lightly draw a line from A to the upper right corner of the illustration "B".

If the illustration is to be reduced to page width, then the height can be scaled from "C" (where the margin intersects the line) to the bottom of the column, "D". If column width, read from E to F. Any intermediate width can also be read (as, for example, at G-H or I-J).

To use the mathematical method, set up a simple ratio-and-proportion equation:

present width	:	*present height*		*reduced width*	:	*reduced height*
5 inches	:	*4 inches*	=	*2.5 inches*	:	*y inches*

Solve the equation: $5y = 4(2.5)$. So the reduced height (y) will be 2 inches.

figuring reductions

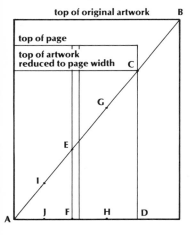

top of original artwork .... B

top of page

top of artwork
reduced to page width ... C

G

E

I

A ... J F H D

writing captions

There is no hard-and-fast rule to guide you in writing captions. Some illustrations need no captions, or only one word. Large maps that are printed on a separate sheet usually have all the information needed distributed about the sheet rather than collected in a single formal caption. Smaller maps, which are incorporated in the text, may or may not be self-explanatory. Perhaps you need a caption reading simply, "Figure 1"; perhaps it should say "Figure 1. Map showing the southeast corner of the northwest part of the Eldorado 15-minute quadrangle, Boulder County, Colorado"; perhaps you need much more. You will have to decide for yourself. Regardless of the length of your caption, be sure to give all necessary credits.

However long or short your captions are, you must supply a typed list of them with your manuscript. You must key each caption to the correct piece of artwork so that the editor cannot possibly mix them up.

If you are an editor, it is your job to see that the author's captions are as clear and complete as possible. Mark the position of each illustration in the manuscript, so that the pertinent text will be printed as closely as possible to the illustration. Before you send the text and artwork to press, check again to be sure that each piece of artwork and each overlay is identified and clearly marked for reduction. You must be especially careful to mark each illustration so that the printer cannot possibly mix them up.

Label each illustration beyond the margin where the label will not be printed. The author should identify himself, his article, and his illustration: "Robert F. Smith—Ross ice sheet—figure 1—green overlay"; the editor can add his journal name and any other information he may need: "Jour Antarctica v 71 n 2 Feb. '78—set in on p. 32". He may also give job numbers, specific ink colors, or other special notations.

color, cost, and care In the past, few geologic illustrations other than complex maps have been printed in color. The reasons are partly cost and partly esthetics.

Color is expensive because it is complicated and difficult to reproduce accurately. (You can spend hundreds of dollars for color separations of a few photographs and still have no printing plates or even a proof.) A good example is minerals, which are often so colorful to the eye but difficult to photograph. The color negative or the color transparency rarely does justice to the subject and the printed illustration has rarely been an accurate representation. Recently, superb color illustrations of minerals have been printed, but to achieve this, close coöperation between photographer, editor, and printer is vital. Great care at all stages is required.

The principal reason that printed color does not resemble the color seen in transparencies or in the minerals themselves lies in the different nature of color as seen, photographed, and printed. The color seen is due to light waves transmitted through your eyes. The color transparency is a step away. If you photograph the three colors—magenta (a red), yellow, and cyan (a blue)—on separate layers of the same film, that shows your eyes the effect of combined translucent colors.

When these colors are translated to printer's ink, a wholly different system of color is used. Ink pigments (none of which may be the same as cyan-yellow-magenta) are added to the printed page as tiny dots of color—yellow-red-blue-black—between and on top of one another. The results of the two methods, although they may be equally pleasing artistically, are seldom identical.

Color photographs are of two types: photos made from color negatives, from which one can have photographic prints made in color or in black and white; and color transparencies, which are "see-through" positives. Either of these can be used to make printing plates. The colors in each must be separated from one another ("color separation") before being used. Ask your editor, if you are using color photographs, which kind of color photo will best fit his production methods.

Hand-drawn color artwork may be sent to the printer separated by the artist, in which case you will submit overlays for the various colors and shades of color to be printed, or unseparated. If the art is unseparated (such as a painting or colored drawing) it must be separated photographically in the same way as a color photograph.

slides for lectures If you are like most scientists you will not make many color photographs or slides with publication in mind. Instead, you will make them in order to illustrate talks and papers to be given to students, colleagues, or the general public. You will, no doubt, want to add other slides to them in order to complete your lecture.

This brings us to another subject, the preparation and presentation of effective lecture slides. There is a great deal to learn about this subject; don't venture into it lightly. It is beyond the scope of this book but, as printed illustrations often become slides without alteration, we want to alert you to danger: rarely will a printed diagram, map, or drawing become a good slide without extra work.

readying the map

Had we but world enough, and time, we could produce maps slowly and beautifully as once we did. But the world is shrinking and time speeding, and today's demands are for maps today. The usefulness of maps is the major reason for the increasing demand. Geologic maps are sophisticated tools, capable of presenting millions of bits of data in an extremely efficient manner.

Today some maps are produced by computer; perhaps most will be in the future. Although our present map-making techniques, computer methods included, are far faster than the lithographic-stone techniques we used until recently, the results are seldom so handsome. Part of this esthetic deterioration is due to a lack of care and knowledge on the part of author, editor, and publisher.

Geologic maps are an efficient method of presenting a great many bits of data—if
- *accurate*
- *completely identified*
- *carefully drawn*
- *thoroughly checked.*
Authors: consult your editor. Editors: consult your printer. Include necessary basic facts. Correlate map, legend, cross-sections, and text.

Before you, as an author, undertake to make a map, it is wise to consult with the editor; if you are an editor, check with the printer. At one state geological survey, for example, each geologist consults with the cartographic supervisor as soon as he receives his assignment, and long before he starts his field work. The supervisor supplies him with the best possible base map to use. Most likely, the base will be one especially compiled for him, probably printed in green on heavy translucent Mylar film. He can draw his map directly on this base, without fear that it will stretch or shrink out of scale. After he has completed his compilation, an ozalid print can be made of the map for checking and editing. When it is ready for the cartographer, the green base can be photographically eliminated, leaving only the author's black lines representing the geological data to be transferred. Or, if needed, the green lines can be retained as check points; being green, they are easily separable from the geologist's black lines.

The map is then transferred to a special film coated with an opaque layer that can be peeled away. The cartographer scribes—cuts this film away—to make as many negatives of the map as there will be different colors and patterns.

This process is the reverse of the lithographic-stone process in that the cartographer now removes what he *does* wish to show, whereas the lithographer etched out what he did *not* wish to show. The scribing process is faster and the material lighter: a scribed map sheet weighs a few grams; an etched stone a few hundred kilograms.

Even when a mapmaker does not know who his publisher will be, he must include certain basic facts on his map (scale, source, title, and the like). Certain others, such as identification numbers, publisher's logo, and distribution outlets, the editor must add.

limits of size

The author should be aware of certain basic limits before he commits himself irrevocably to a certain style or size of map. Large maps, besides being unhandy, do not endure; they are often folded wrongly so many times that soon they disintegrate in a mass of crinkles. In addition to unwieldiness, there is a limit to paper size and to press size. One major publisher of geologic maps, for example, has a press that will accommodate no more than 42×58 inches. Even that size map is so large that one must crawl across it to see detail at the center; it may serve in library or classroom, but is almost unmanageable in the field.

Although the author seldom has control of the kind of paper used, editors do—and they should insist on long-lasting, acid-free paper in durable, acid-free pockets. Paper with substantial content of sulfur disintegrates quickly on the library shelves.

symbols   Commonly used symbols for topographic maps are included in the U.S. Geological Survey's free pamphlet, *Topographic maps*. Very valuable, but hard to obtain, is the Survey's in-house release by its Topographic Division entitled "Symbols for standard topographic maps published at the scale 1:63360 and larger to be scribed at 1:24000–1:48000", which gives an extensive list of symbols, together with approved cartographic line weights.

There is no universally recognized standard for geologic map symbols (including contacts, lithologic types, structural relationships). There are many reasons why there is none: long-standing company or publication procedures; availability of symbols on stencils, lettering sets, stick-on drafting overlays, and the like; different purposes and audiences for intended maps; and varied mapping scale and detail, and the complexity of geology itself.

Here are some suggestions for choosing the most appropriate symbols for your map:

1. Review the U.S. Geological Survey system for geologic map symbols. They have a printed but unpublished sheet of revised symbols, and map symbols are listed in references cited in the section "Reference shelf".

2. Review the Unesco *International legend for hydrogeological maps* (1970; Cook, Hammond & Kell Ltd., England). Other Unesco publications may also help.

3. Review other maps drawn to similar scales—if any exist—to show features similar to yours in other areas.

4. Most important, include a complete legend or explanation that defines all the symbols you use in your map, including the significance of various tints, lettering styles, and other overprints used.

check list   The map, legend or explanation, and cross-sections should be checked for

• Completeness. Are all units labeled? Are all formations in legend and cross-section also on map? Are all geographic locations mentioned in text shown on map?

• Correctness. Are there errors in plotting? Are names spelled correctly? Does the spelling on the map agree with that in the text? Are the locations as described in text and shown on cross-sections in agreement with the map? Are numbers correct? Do names and ages in legend agree with map and text? Are all symbols in legend on map and those on map in legend? Do colors or patterns in map and sections match legend? Do dips as drawn in cross-section agree with the map?

• A concise, yet complete, title including subject and location. Include the state, possibly the county, and perhaps the country, as "Geologic map of Kern County, California", or "Map of the tectonic features of the United States".

• Author's or compiler's name; contributors and sources of data or bases.

• Scale, preferably both graphic and numerical, in metric and English units.

• Contour interval and datum.

• Inclusive dates of field work or compilation; date of publication or copyright.

• Name of publisher and place of publication; name and address of distributor, if needed.

• Series and sheet identification, if more than one sheet, as "Economic Mineral Investigation 3—sheet 2".

- Arrows showing true and magnetic north; declination.
- Sponsors, draftsmen, cartographers (so labeled).
- Geographic reference points and grids (township-range grid; longitude-latitude; Army Map Service grid).
- Lines of cross-sections.
- Rapid identification in upper right-hand corner, as "Map Sheet 54, Dubuque quadrangle".
- Color or pattern block and explanation of all units on map (most maps will include age designations and abbreviated lithology).
- Explanation of symbols.
- Title, explaining what they are sections of.
- Horizontal scale, if different from the map or separated from it.
- Vertical scale or vertical exaggeration.
- Orientation.

For United States national map-accuracy standards, write to the U.S. Geological Survey, Reston, Va., 22092. A published map meeting these requirements should say in the legend that "This map complies with national map-accuracy standards."

rules for names

- *Rules for naming things, times, and places help us understand what others mean.*
- *Lithostratigraphic, biostratigraphic, chronostratigraphic, lithologic, and geochronologic units should be kept separate, and identified when used.*
- *The geologic column allows correlations between time and time-rock units.*
- *Conventions for use of geologic, geographic, and taxonomic units are established by special committees.*

Things, times, and places form the basic framework around which most geological reports are written. Over the years, certain conventions have been adopted and adapted for effective communication. That there is no authoritative final word is quite clear from the deliberations of stratigraphers, paleontologists, mineralogists, zoologists, botanists, and others who concern themselves with details like these. National and international commissions and committees deal with nomenclature, priorities, and the like.

The commentary below concerning various types of names and their usage is freely adapted from various sources, especially the U.S. Geological Survey and the International Subcommission on Stratigraphic Classification.

Stratigraphic classification allows for systematic organization of rock strata into units with reference to any of several characteristics or properties they possess, especially lithologic character, fossil content, age and time relationships, seismic properties, magnetic reversals, electric-log characteristics, mineral assemblages, lithogenesis, and environments of deposition or formation.

Lithozones, biozones, chronozones, mineral zones, and others, usually designate minor stratigraphic intervals in respective classification categories. When used formally as a named unit, the first letter of the zone name is capitalized.

Lithostratigraphic units consist mainly of certain lithologic types or other significantly unifying lithologic features. Examples are the Dundee Limestone of Middle Devonian age in the Michigan basin, and the Flathead Sandstone of Cambrian age in Montana and Wyoming. The formation is the fundamental unit of lithostratigraphic classification, and may be subdivided into members and beds. Two or more associated formations having significant lithologic features in common may be included in a group, such as the Cisco Group.

Biostratigraphic units are designated by fossil content or paleontologic character that differentiates them from adjacent units. Examples are the *Heterostegina* Assemblage Zone of the Gulf Coast area, and the *Cardioceras cordatum* Range Zone.

Chronostratigraphic units include rocks formed during some specified interval of geologic time. Examples are the Paleozoic Erathem, Silurian System, Middle Silurian Series, and Tonawandan Stage. Corresponding geochronologic terms, representing the time during which those units were formed, are era, period, epoch, and age. Consequently, rocks of the Silurian System were deposited in the Silurian Period of the Paleozoic Era.

The stratigraphic column in this book is organized to allow a generalized correlation between measured time and the occurrence of time-rock units.

guides to usage

A series of guides to U.S. stratigraphic usage has been prepared by the American Commission on Stratigraphic Nomenclature. The latest revision is available from the American Association of Petroleum Geologists (1970, Tulsa; 22 p.) under the title *Code of stratigraphic nomenclature*; it formulates an explicit statement of principles and practices for classifying and naming stratigraphic units.

Here are a few of the conventions recommended by the code. All words used in names of formal rock-stratigraphic (lithostratigraphic) units should have initial capitals, such as Ash Creek Group, Chinle Formation, Kirtland

Shale, Church Rock Member, Sonsela Sandstone Bed; however, references to informal units, such as an unnamed sandstone bed in the Chinle Formation, do not follow that rule: Chinle sandstone bed. Capitalization of formal and informal units of time (geochronologic units) and rock-time (chronostratigraphic units) follows similar conventions, thus: Paleozoic Era, Paleozoic Erathem, Devonian Period, Cenomanian Stage, Cenomanian Age, *but* Devonian time, Devonian age, and Paleozoic age (these last are not specific time units). Rules for formal zone names are the same, except that the italicized (or underscored) species name of a plant or animal is never capitalized, e.g., *Bulimina excavata* Concurrent-range Zone.

Lower, Middle, and Upper are formal series subdivisions of a system, and Early, Middle, and Late are the corresponding formal (and therefore capitalized) *time* terms. For example, Upper Cretaceous rocks were deposited in Late Cretaceous time.

Informal chronostratigraphic terms, lower, middle, and upper, and the corresponding informal time terms, early, middle, and late, are used as subdivisions of eras, formal series of the Tertiary, such as lower Pliocene or early Pliocene, and provincial series, such as lower Atokan or early Atokan. These adjectives are not capitalized.

Proposed lithostratigraphic units should be described and defined clearly, so that others can recognize them. The intent to introduce a new name and the important factors that led to discrimination of the unit should be clearly stated. The definition should give the geographic or other feature from which the name is taken, and the specific location of one or more representative sections near the geographic feature. Specific reference to location in section, township, and range, or other land divisions should be included. The thickness, lithology, color, and age of the unit should be given.

U.S.G.S. Bulletins
U.S. Geological Survey Bulletins 896, 1200, and 1350 provide basic information on recognized geologic units in the United States through 1967.

If you have doubts about specific place names, check with the Board of Geographic Names (c/o U.S. Geological Survey, Reston, Va., 22092).

Fossil names should adhere to conventions of the International Rules of Zoological Nomenclature and the International Code of Botanical Nomenclature.

This column shows a general correlation between absolute time and the occurrence of time-rock units. Dates at system or series boundaries are in millions of years, and are, of course, approximations. The smaller divisions are stage names except for the Cambrian series listed; the level of the North American Cenozoic subdivisions is unsettled. The left column is the European section; the right one, the North American section. Important orogenic climaxes are in brackets.

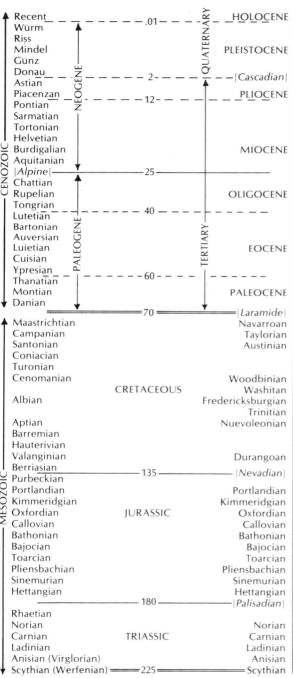

CENOZOIC — NEOGENE / PALEOGENE (European) — QUATERNARY / TERTIARY (North American)

European	(m.y.)	North American
Recent	.01	HOLOCENE
Würm		
Riss		
Mindel		PLEISTOCENE
Günz		
Donau	2	|Cascadian|
Astian		PLIOCENE
Piacenzan	12	
Pontian		
Sarmatian		
Tortonian		
Helvetian		
Burdigalian		MIOCENE
Aquitanian		
|Alpine|	25	
Chattian		
Rupelian		OLIGOCENE
Tongrian	40	
Lutetian		
Bartonian		
Auversian		
Luietian		EOCENE
Cuisian		
Ypresian	60	
Thanatian		
Montian		PALEOCENE
Danian		
	70	|Laramide|

MESOZOIC

European	(m.y.)	North American
Maastrichtian		Navarroan
Campanian		Taylorian
Santonian		Austinian
Coniacian		
Turonian		
Cenomanian		Woodbinian
CRETACEOUS		Washitan
Albian		Fredericksburgian
		Trinitian
Aptian		Nuevoleonian
Barremian		
Hauterivian		
Valanginian		Durangoan
Berriasian	135	
Purbeckian		|Nevadian|
Portlandian		Portlandian
Kimmeridgian		Kimmeridgian
Oxfordian	JURASSIC	Oxfordian
Callovian		Callovian
Bathonian		Bathonian
Bajocian		Bajocian
Toarcian		Toarcian
Pliensbachian		Pliensbachian
Sinemurian		Sinemurian
Hettangian		Hettangian
	180	|Palisadian|
Rhaetian		
Norian		Norian
Carnian	TRIASSIC	Carnian
Ladinian		Ladinian
Anisian (Virglorian)		Anisian
Scythian (Werfenian)	225	Scythian

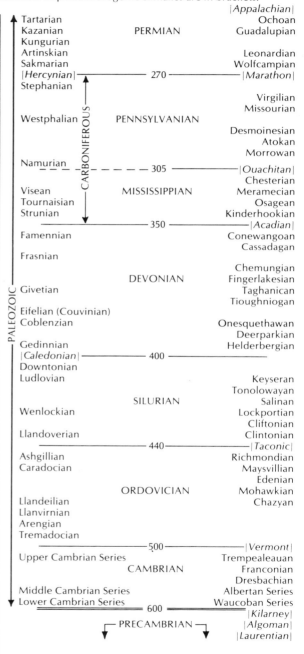

PALEOZOIC — CARBONIFEROUS

European	(m.y.)	North American
		|Appalachian|
Tartarian		Ochoan
Kazanian	PERMIAN	Guadalupian
Kungurian		
Artinskian		Leonardian
Sakmarian		Wolfcampian
|Hercynian|	270	|Marathon|
Stephanian		
		Virgilian
Westphalian	PENNSYLVANIAN	Missourian
		Desmoinesian
		Atokan
Namurian	305	Morrowan
		|Ouachitan|
Visean		Chesterian
Tournaisian	MISSISSIPPIAN	Meramecian
		Osagean
Strunian		Kinderhookian
	350	|Acadian|
Famennian		Conewangoan
		Cassadagan
Frasnian		
		Chemungian
		Fingerlakesian
	DEVONIAN	Taghanican
Givetian		Tioughniogan
Eifelian (Couvinian)		
Coblenzian		Onesquethawan
		Deerparkian
Gedinnian		Helderbergian
|Caledonian|	400	
Downtonian		
Ludlovian		Keyseran
		Tonolowayan
	SILURIAN	Salinan
Wenlockian		Lockportian
		Cliftonian
Llandoverian		Clintonian
	440	|Taconic|
Ashgillian		Richmondian
Caradocian		Maysvillian
		Edenian
	ORDOVICIAN	Mohawkian
Llandeilian		Chazyan
Llanvirnian		
Arengian		
Tremadocian		
	500	|Vermont|
Upper Cambrian Series		Trempealeauan
	CAMBRIAN	Franconian
		Dresbachian
Middle Cambrian Series		Albertan Series
Lower Cambrian Series		Waucoban Series
	600	
		|Kilarney|
PRECAMBRIAN		|Algoman|
		|Laurentian|

on citing sources

When you write a technical paper you must identify information you have garnered from other sources, and you must identify the sources themselves. This accountability sets professionals apart and gives credit where credit is due.

You can avoid trouble for yourself, authors, editors, reviewers, and publishers by using common sense: find whether a particular citation style is required. Most journals do require a particular style, and those styles vary widely, so check the style book if there is one. If there is no style book, study examples in the journal itself. If all else fails, devise a style suitable to the subject matter and use it consistently.

Some rules for reference citations seem complex but most are necessary. (The fact that much editorial time is spent cleaning up reference lists shows how poorly the rules are understood—and how much editors value them.)

If you are responsible for setting a citation style, aim for simplicity and utility. That will help your readers—and it is for their benefit that you include the information to begin with.

Note that the typographic style of citations may take several forms such as footnotes and parenthetical notes (Hess, 1962, for example). The editorial style involves alphabetization, chronology, abbreviations, and the order of elements within the citation.

The adjacent check list shows the kinds of information that should be considered for inclusion, regardless of the exact style. (This sequence is not necessarily the best for all purposes.)

Examples of a system for citation are evident in almost any long bibliography.

A little care can make your contribution more readily citable by others. Remember that increasing use is being made of machine-oriented indexing and searching techniques for the literature; it helps to keep your titles short and make sure they tell what the contribution is about. Most words in the title should be those that when indexed will help a reader searching the literature by key-word indexes (this is true whether the searching is by a person or by a computer program).

It has been estimated that 15,000 serial publications include papers pertinent to geology. GeoRef, a system whose data base is used to compile the *Bibliography & index of geology*, includes about 3,000 serial titles; about 2,000 of these include about 95 per cent of the world's geological literature.

A recent trend among editors is to avoid the confusion that arises through use of journal and other serial abbreviations by asking authors to spell out the titles in their reference sections. This trend is to be lauded, particularly for foreign and otherwise uncommon sources of literature. The space saved by abbreviations, many editors contend, is more than overbalanced by frustration to readers and by the time that editors spend in checking abbreviations and making corrections.

reference elements
- *Author's surname and given name or initials—used the way the author does*
- *or the editor's surname and given name.*
- *Publication date (the year usually suffices).*
- *Title.*
- *Special category of publication, such as abstract, edited work, editorial, photograph.*
- *Publication (series) title, such as journal, symposium issue, field guide.*
- *Volume number (and part, and issue number, if the pages are not numbered consecutively).*
- *Publisher.*
- *Address of publisher (especially for non-serial publications).*
- *Pagination (total if entire work, or that included in contribution if it is only part of a work).*
- *Information that will enable a person to find the work or get a copy of it, if it is "unpublished".*

abstracting
the essence

The most important part of a paper may be its abstract. Effective abstracts
- *are concise*
- *summarize conclusions and recommendations*
- *are amenable to computer storage and retrieval.*

In terms of number of readers, an abstract is easily the most essential part of a technical paper. It is like the bouillon cube and unlike the bouillon-cube's wrapper: the true essence is not in the list of ingredients.

Two views of abstracts follow. The first is Kenneth K. Landes' 1966 classic, "A Scrutiny of the Abstract, II":

A partial biography of the writer is given. The inadequate abstract is discussed. What should be covered by an abstract is considered. The importance of the abstract is described. Dictionary definitions of "abstract" are quoted. At the conclusion a revised abstract is presented.

For many years I have been annoyed by the inadequate abstract. This became acute while I was serving a term as editor of the *Bulletin* of the American Association of Petroleum Geologists. In addition to returning manuscripts to authors for rewriting of abstracts, I also took 30 minutes in which to lower my ire by writing "A Scrutiny of the Abstract". This little squib has had a fantastic distribution. If only one of my scientific outpourings would do as well! Now the editorial board of the Association has requested a revision. This is it.

The inadequate abstract is illustrated at the top of the page. The passive voice is positively screaming at the reader! It is an outline, with each item in the outline expanded into a sentence. The reader is told what the paper is about, but not what it contributes. Such abstracts are merely overgrown titles. They are produced by writers who are either (**1**) beginners, (**2**) lazy, or (**3**) have not written the paper yet.

To many writers the preparation of an abstract is an unwanted chore required at the last minute by an editor or insisted upon even before the paper has been written by a deadline-bedeviled program chairman. However, in terms of market reached, the abstract is *the most important part of the paper*. For every individual who reads or listens to your entire paper, from 10 to 500 will read the abstract.

If you are presenting a paper before a learned society, the abstract alone may appear in a pre-convention issue of the society journal as well as in the convention program; it may also be run by trade journals. The abstract which accompanies a published paper will most certainly reappear in abstract journals in various languages, and perhaps in company internal circulars as well. It is much better to please than to antagonize this great audience. Papers written for oral presentation should be *completed prior to the deadline for the abstract*, so that the abstract can be prepared from the written paper and not from raw ideas gestating in the writer's mind.

My dictionary describes an abstract as "a summary of a statement, document, speech, etc. . . ." and that which *concentrates in itself the essential information* of a paper or article. The definition I prefer has been set in italics. May all writers learn the art (it is not easy) of preparing an abstract containing the *essential information* in their compositions. With this goal in mind, I append an abstract that should be an improvement over the one appearing at the beginning of this discussion.

revised abstract   The abstract is of utmost importance, for it is read by 10 to 500 times more people than hear or read the entire article. It should not be a mere recital of the subjects covered. Expressions such as "is discussed" and "is described" should *never* be included! The abstract should be a condensation and concentration of the *essential information* in the paper.

The second view is illustrated by excerpts from "Standards for writing abstracts" by B.H. Weil:

the abstract defined An abstract, as defined here, is an abbreviated, accurate representation of a document. The following recommendations are made for the guidance of authors and editors, so that abstracts in primary documents may be both helpful to their readers and reproducible with little or no change in secondary publications and services.

Make the abstract as informative as the document will permit, so that readers may decide whether they need to read the entire document. State the purpose, methods, results, and conclusions presented in the document, either in that order or with initial emphasis on findings.

For various reasons, it is desirable that the author write an abstract that the secondary services can reproduce with little or no change. These reasons include the economic pressures on the secondary services caused by continuing increases in the volume of scholarly publication; the need for greater promptness on the part of the secondary services in publishing information about the primary literature; and the growing value of good authors' abstracts in computerized full-text searching for alerting and information retrieval.

In the proposed standard the term *abstract* signifies an abbreviated, accurate representation of a document without added interpretation or criticism and without distinction as to who wrote the abstract. Thus, an abstract differs from a brief *review* of a document in that, while a review often takes on much of the character of an informative or informative-indicative abstract, its writer is expected to include suitable criticism and interpretation. While the word *synopsis* was formerly used to denote a résumé prepared by the author, as distinct from an *abstract* (condensation) prepared by some other person, this distinction no longer *has* real meaning.

abstract to inform An abstract should be as *informative* as is permitted by the type and style of the document; that is, it should present as much as possible of the quantitative and/or qualitative information contained in the document. (Stringencies in publication economics may be governing factors, but they do not in themselves set standards for the quality of an abstract.) Informative abstracts are especially desirable for texts describing experimental work and documents devoted to a single theme. However, some discursive or lengthy texts, such as broad overviews, review papers, and entire monographs, may permit the preparation of an abstract that is only an *indicative* or descriptive guide to the type of document and what it is about. A combined *informative-indicative* abstract must often be prepared when limitations on the length of the abstract or the type and style of the document make it necessary to confine informative statements to the primary elements of the document and to relegate other aspects to indicative statements.

Abstracts should not be confused with the related, but distinct, terms *annotation, extract,* and *summary.* An *annotation* is a note added to the title or other bibliographic information of a document by way of comment or explanation. An *extract* signifies one or more portions of a document selected to represent the whole. A *summary* is a restatement within a document (usually at the end) of its salient findings and conclusions, and is intended to complete the orientation of a reader who has studied the preceding text. Because other vital portions of the document (e.g., purpose, methods) are not usually condensed into this summary, the term should not be used synonymously with "abstract"; i.e., an abstract as defined above should not be called a summary.

A well-prepared abstract enables readers to identify the basic content of a document quickly and accurately, to determine its relevance to their interests, and thus to decide whether they need to read the document in its entirety. Readers for whom the document is of fringe interest often obtain enough information from the abstract to make their reading of the whole document unnecessary. Therefore, every primary document should include a good abstract. Secondary publications and services that provide bibliographic citations of pertinent documents should also include good abstracts if at all possible.

where to use abstracts   The following recommendations are for authors and editors of specific documents and publications.

Include an abstract with every formal item, such as research, methods, and theoretical papers; speculative and hortatory articles; essays; discussions; and review articles. Notes, short communications, editorials, and Letters to the Editor that have substantial technical or scholarly content should also have brief abstracts.

Include an abstract in every separately published report, pamphlet, or thesis. If a report must receive a government security classification, it is highly desirable for documentation purposes "that the abstract be unclassified if the results obtained can be reported in the abstract only in general terms."

A single abstract may suffice in a book or monograph that deals with a homogeneous subject. However, a separate abstract is also necessary for each chapter if the volume covers many different topics or is a combination of articles by different authors (e.g., the proceedings of a meeting or symposium).

An abstract is now included in every United States patent, and the United States Patent Office has established its own guidelines for them.

secondary services   Secondary publications and services can often make verbatim use of the abstracts provided in primary documents if these abstracts have been carefully prepared and are free from copyright restrictions. Such authors' abstracts can also provide suitable bases for the secondary service that orients its abstracts to a group of users different from those envisioned by the authors. A completely new abstract usually needs to be written only when brief, subordinated phases of a document are all that fall within the scope of a secondary publication.

Since an abstract must be intelligible to a knowledgeable reader without reference to the document, make the abstract self-contained. Retain the basic information and tone (balance, emphasis) of the original document. Be as concise as possible while still fulfilling requirements as to content, but do not be cryptic or obscure. Cite background information sparingly if at all. Do not include information or claims not contained in the document itself.

completeness and accuracy   For most papers and portions of monographs, an abstract of fewer than 250 words will be adequate. For notes and short communications, fewer than 100 words should suffice. Editorials and Letters to the Editor often will require only a single-sentence abstract. For long documents such as reports and theses, an abstract generally should not exceed 500 words and preferably should appear on a single page.

Begin the abstract with a topic sentence that is a central statement of the document's major thesis, but avoid repeating the words of the document's title if that is nearby.

In abstracts specifically written or modified for secondary use, state the type of the document early in the abstract when this is not evident from the title or publisher of the document or will not be clear from the remainder of the abstract. Explain either the author's treatment of the subject or the nature of the document, e.g., theoretical treatment, case history, state-of-the-art report, historical review, report of original research, Letter to the Editor, literature survey.

Write a short abstract as a single, unified paragraph, but use more than one paragraph for long abstracts, e.g., those in reports and theses. Write the abstract in complete sentences, and use transitional words and phrases for coherence.

active verbs   Use verbs in the active voice whenever possible; they contribute to clear, brief, forceful writing. The passive voice, however, may be used for indicative statements and even for informative statements in which the receiver of the action should be stressed.

Avoid unfamiliar terms, acronyms, abbreviations, or symbols; or define them the first time they occur in the abstract.

Include short tables, equations, structural formulas, and diagrams only when necessary for brevity and clarity.

judgment by peers

Why technical reviews?

Every manuscript submitted for publication should represent the author's best efforts. Ideally, this means that it has been written and rewritten, set aside for a cooling period, and again rewritten and polished. Even if all this is done, however, and the manuscript represents the best job in reasoning, exposition and over-all organization of which the author is capable, every paper will still benefit from technical review by others. The author is too close to his work; a fresh, objective look at it by someone else is essential in order to spot errors in fact or reasoning, inconsistencies, or even generally poor presentation that obscures what the author has tried to convey.

Technical review may be sought from a colleague, either during or after preparation of a manuscript. Advice from such sources is helpful, if not sometimes indispensable, in the formulation of ideas, but it is seldom sufficient in itself. This is primarily because colleagues usually already know a good deal about the author's project from earlier conversations, and hence may fall into the same traps as the author. Moreover, colleagues' reviews are apt to be cursory, and may also suffer from an unconscious attitude of "I'll scratch your back if you'll scratch mine." In summary, formal or informal reviews by the author's co-workers, or by his professors or superiors, though highly recommended, should not take the place of objective reviews by outsiders.

Virtually any large research organization that issues its own publication, as well as any journal that publishes formal papers submitted by the technical and scientific "public", maintains some sort of technical review system for manuscripts. The purpose is to provide independent advice to the editor as to the technical quality of all submitted manuscripts, to aid him in selection or rejection of manuscripts within the needs and policies of his publication, and to aid authors in improving their presentations. A few journal editors, particularly of small, highly specialized journals, can do all these things themselves, and so can dispense with outside reviewers. In all other cases, however, technical review is considered an essential part of the writing-publishing process.

Manuscripts, as they are submitted to editors for publication, range from very poor to excellent. Not surprisingly, their over-all quality, both in thought and expression, is directly related to the quality and quantity of the technical review they have already undergone. There are exceptions, of course, but there are usually wide differences between manuscripts from independent geologists or from small, understaffed and overworked college faculties and those from the larger institutions that can maintain elaborate systems for manuscript processing and in-house review.

These differences are well known to most editors. The knowledge may lead them to bear down harder on the paper that is likely to have had inadequate prior review. Among most editors, surely, this possible difference in treatment is emphatically not due to prejudice. The evidence of painstaking field work or of brilliant gems of deduction is quite as likely to come from the mind of a lone worker as it is from one in a great institution. The difference is that the lone worker's contribution may need more review and polish to bring out its qualities.

a typical system   Technical-review systems are almost as numerous and varied as the publishing houses that have adopted them. Only one such system is described here, one reasonably comparable to those followed by many earth-science journals. (For another discussion, see Frank T. Manheim's "Referees and the publications crisis", in E⊕S, v. 54, May 1973.)

In this system, a full-time editor is responsible for all of his society's publication program. In the present context, however, his chief responsibility is to make final decisions on the technical and scientific quality of all papers and books published by the society. No one geologist can possibly know all he needs to know about all the earth sciences in order to make the required judgments. Instead, the editor relies heavily on a group of associate editors and on technical reviewers chosen by them.

When a new manuscript has been received, the editor scans it just enough to decide which associate editor is most likely to be able to handle it. These associate editors, all volunteers, are chosen to provide as broad a spectrum as possible of topical and geographic knowledge of the specialties that are collectively known as earth science.

choosing reviewers

The associate editor is asked in advance, usually by phone, to make sure that he has time and the requisite knowledge of the subject matter to deal with the new manuscript promptly. If he agrees, he selects one or more technical reviewers who are broadly or specifically knowledgeable in the subject matter, and who are able and willing to do a chore on behalf of the society and of the author.

The technical reviewers send their advice and recommendations to the associate editor, on printed forms, as written commentaries, or as marks and notes throughout the text. The associate editor then reviews the manuscript and the reviewers' comments and advises the editor of the combined results. The editor must then make the final judgment and transmit it to the author.

The associate editor and the technical reviewers are asked to read a paper, and advise on its acceptability, from the standpoint of scientific soundness, originality, breadth of interest, and length. In judging length, some compromise is usually desirable between the author's need to tell a complete story and the publisher's need to conserve space.

Organization and style are also critically examined by the technical reviewers, because no research results, no matter how excellent, will be read or understood if they are poorly expressed. In general, however, matters of style and expression are considered less important at the review stage than are the other qualities just listed. Generally, they can be improved later while the manuscript is being polished and prepared for the printer. The chief object of the technical review process is to help the editor decide whether to accept a given manuscript, not to rule on details of its presentation.

pursuit of excellence

The review process is not intended to stifle the author scientifically or to force him to agree with preconceived notions held by the technical critics. No reputable publisher wants to act as a scientific censor. However, all publishers seek scientific excellence, neatly and clearly presented. The purpose of the review process is to find such excellence and to help the author bring out the best in his work for the ultimate advancement of the science.

A tiny fraction of submitted manuscripts come through the review gantlet unscathed, and are recommended for publication virtually without change. These are turned over to the managing editor, whose staff will prepare the manuscript for the printer and see that it is published.

A significant percentage of submitted manuscripts is rejected for publication after review. Some outright rejections are based on judgments that the subject matter belongs more properly in some other publication. Other rejections are based on a variety of reasons; most have to do with the factors that the critics have been asked to consider, but they include such other reasons such as costs of illustrations or lengths that are too great. Editors realize that, no matter how firmly rejected, most manuscripts will inevitably come back to them or go to other editors in some form or another. This is good, for almost every piece of research contains some elements of truth that deserve publication somewhere.

The vast majority of reviewed manuscripts are returned to the authors with requests for revision, which range from minor to major in scope. A request for major revision may require what amounts to a completely new paper. In such cases, the editor may put the new version through the review process again, often using different reviewers than those who read the earlier version. In most cases, however, the editor reads enough of the revised version to satisfy himself that the author has made a conscientious effort to follow the critics' advice. If he has, the manuscript is accepted for publication and put in the mill.

author vs. reviewer   Authors react to criticism of their manuscripts in a variety of ways, depending on their personalities, their relations with their critics, and their maturity. Some welcome help, or at least acquiesce gracefully; others fly into a rage and see ignorance, if not foul play, in every mark made by their sworn enemies, the technical reviewers. Many an author has had to drop his work and at least walk around the block in an effort to cool his temper and his hurt feelings after his manuscript was returned from review.

Authors such as those just described seldom believe it until they become critics themselves, but the fact is that with rare exception critics are people of good will, genuinely trying to help the author. Criticism is at best a thankless job, done by people who would much rather pursue their own research than review manuscripts by others. Rarely, a critic may run across a manuscript that contains some gem of new thought within his own specialty. If this occurs, he is grateful to the author and for this contribution to his own awareness. More often, the critic is simply an unpaid volunteer, doing a job out of loyalty to science, to the publishing society or institution, and to the author.

The author, then, should approach the critics' comments on his manuscript with an open, cool mind. He should force himself, if necessary, to believe that the critic is on his side and that every comment, no matter how sarcastic or inept, deserves his thorough and objective consideration.

Many critical comments, or interlineations on the manuscript, seem at first glance to be so wrong as to imply gross carelessness, if not downright stupidity, on the part of the critic. These suspicions must be fought down or ignored. Instead, the author must assume that the more "stupid" a critic's comment, the more the author must study his original words. Surely something in the expression, the facts presented, or the reasoning led the critic astray, and caused him to make the "stupid" comment or mark.

fairness—and tact   Charges of prejudice, of self-dealing, and of other forms of foul play come quickly to the minds of a few authors when they receive their manuscripts back from the review process. Dealing with such charges, fairly to all concerned, constitutes one of the more delicate responsibilities of a journal editor. With long experience, the editor may come to identify an author as a chronic victim of "persecution"; future manuscript offerings from that author will probably be unwelcome and will get short shrift. In more cases than not, cries of "foul" are based on personal conflicts or jealousies, or on real or supposed domination of some area of knowledge by the Establishment, usually a particular organization or school of thought.

a second opinion Methods of handling such charges differ widely among various editors, and journals. It is sound policy to accept charges of "foul" in good faith: the author should be offered another hearing by a new set of reviewers, and should be given the opportunity to nominate one or more of the reviewers himself. This is a dangerous practice, as no downtrodden author is likely to choose an unsympathetic reviewer, but it may be advisable in the interest of fairness.

the final arbiter Regardless of how such charges are handled, the editor must be the final arbiter. In this difficult task he can perhaps take comfort from the realization that the author of an individual paper will get his just deserts of fame or blame to far greater degree than will the publication medium. This observation should, in fact, be ever uppermost in every author's mind throughout his tribulations in revising his manuscript after technical review.

editing and proofing

Today a completed manuscript can be printed a few weeks after acceptance: edited to a widely accepted style, composed by electronic machines that set thousands of characters per minute, printed by presses that complete a hundred impressions of 32 pages each per minute, and also bound as it comes off the press.

The time is not far off when each earth scientist may be his own editor and typesetter. Automatic machines read directly from typewriter copy, and others permit editing of copy projected on a television-like screen. Some day soon authors may be expected to send manuscripts straight to the machine.

It is increasingly important that each author take responsibility for making his manuscript complete, accurate, and well written. Whether your article is to be printed from old-fashioned handset type or with the aid of an automatic scanning machine, you must submit the best copy you possibly can. You should see that the manuscript conforms to the style of the journal to which you are submitting it; that it has been reviewed by your colleagues; and that there is adequate supporting material (illustrations, tables, and maps) to add clarity and interest.

In addition to review by your own colleagues, your paper will undergo a technical review and editing by one or more scientists familiar with your subject.

At one professional society, your paper will be reviewed by one or more associate editors familiar with the subject. It is then given to the copy editor to

1. Identify and number all parts of the author's copy (manuscript pages, illustrations, tables, foldouts, etc.).
2. Check the manuscript for organization, grammar, and punctuation.
3. Read the text again to see that the title is specific, concise, and includes a locality (if appropriate); that the abstract is short, informative, and specific; that the body of the text is concise and style is consistent.
4. Query the author on points of clarity, need for condensing, and, if necessary, suggest a better choice of words.
5. Check cited references in the text against a bibliographical list, to make sure there are no omissions or superfluous entries.
6. Mark the manuscript for type sizes and similar specifications.
7. Make sure that the order of headings and subheadings is logical and consistent.
8. Proportion illustrations, check spelling, drafting, and captions.
9. Return the manuscript to the author for final approval.
10. Reread the manuscript when it is returned, to incorporate any changes or additions.

Editing is intended to be flexible and to enhance the author's expression of his scientific information. Style rules and regulations of the publisher should not be so rigid that they cannot bend. Both are designed as patterns for you to follow; the tailoring is your own.

Until the time comes that there is no editor between you and the reader, the author's responsibility for the printing process ends with his approval of the edited manuscript.

It is up to the editor to mark the manuscript for the typesetter and to marshal the proofs through the press. True, the author may be called on to read one or more proofs, but the editor will also read them, and the editor will relay the author's notes to the printer.

clarity The first aim in editing manuscript copy or marking proofs is clarity. At every point the typesetter must understand exactly what author and editor want done, and he must be able to understand it without erasing material accidentally elided, deciphering careless handwriting, or puzzling out the meaning from the context. Copy and proofs must be marked so that the type could be set by a typesetter who knows no English whatever.

Manuscript copy and galley or page proofs are marked in slightly different ways. However, the rule of common sense will cover all cases.

Manuscripts must be double or triple spaced so that a change can be inserted where it is needed. Manuscript copy should read in a continuous line, without distracting detours from the middle of the page to the margin and back again. Such detours invite errors by the typesetter. Also, margins should be reserved for mechanical instructions to the printer.

The symbols listed here are all generally accepted forms, but a few that are commonly used have been omitted because of duplication or ambiguity—hence only one form is shown. (We see no reason to encourage use of, say, a half dozen variations of the deletion symbol.)

Try to use a different color pencil for your marks and corrections than has been used by anyone else who has read the copy. Each person marking a manuscript—author, editor, reviewer, typesetter—should indicate the color he has used by his initials.

In general, anything not to be set in type should be circled: for example, the editor's query to the author, or instructions to the printer.

copyediting

comma	Standard symbols when used wisely help the type-
semicolon	setter others, too. It may help to remember this
colon	quotation
quotation mark	Type is set letter by letter and space by space, so
apostrophe	each letter or space must be clear. Don't forget.
hyphen	Well=edited copy is marked to show the length of
1-em dash	each dash ¦ as in this example: the printer Johann
1-en dash	Gutenberg, 1400¦1468.
question mark	Doesn't that seem logical?
exclamation point	Of course!

parentheses	Parentheses (as in this example) are often necessary;
brackets	brackets [editor's notes] are somewhat less common,
braces	and braces {like these} are rarer still.
delete	Editors must mark deletions precisely to prevent
delete, close up to normal	error, bearing in mind that someone must decide what
delete, close up entirely	to do with the space left over.
reduce space	Spacing can be wrong in several ways: here too
add space	much, here too little, and in the next line it is
even spacing	both irregular and uneven as you see.
add space	Sometimes the spacing between lines is too little;
close up	sometimes it is too much.
indent	Indicate paragraph indentation where necessary.
flush	Or show that you don't want normal indentation.
	If there can be any doubt, mark it "flush".
run in	Flush Or "run in" if that is appropriate.
start new line	You can do this: Start a new line.
run over	You can ask for a letter (or letters) to be run o-
	ver to the next line, or to be run back to the pr
run back	eceding line.
1-em indent	Indentations can be marked like this.
2-em indent	Or this.
flush left	This way for flush left,
flush right	this for flush right,
center	and this for center.

align type	Sometimes type is un~ven and should be marked.
ragged right	You can ask for a ragged margin (mark with a wavy line).
transpose letters	Sometimes letters are transposed, or the word
transpose words	order wrong is. Complicated transpositions can be
followed numbered order	marked as in "letters and words".
author, please clarify	If only the author can clarify a point (does he really mean to spell it Psmith?), ask him.

It is customary to circle figures and words, as in

7; seven (seven) and (7), if you mean "do the opposite" -- but circling also means "do not set" as in the next paragraph.

Marking type specifications calls for both brevity and inaccuracy; it takes too long to write "Set in 9-point Optima, on an 11-point slug, 26 picas wide".

The most common form of a typeface is called

roman	(sometimes inaccurately) roman. (rom)
boldface	Boldface is used for sideheads in this book. Scien-
italic	tific names are set in italic, as Atrypa reticularis.
SMALL CAPS	Other special uses call for small capitals, as here;
CAPS AND SMALL CAPS	capitals and small capitals (referred to as "caps
BOLD CAPS	and small caps") and for bold caps. (BC)
lower case	Capitals can be marked to be set in lower case,
CAPITALS	lower case can be marked for capitals, and so on.

superscript; subscript	Subscripts and superscripts, as in $^{14}_{12}C$, should be marked clearly, as should anything that the typesetter might take to be an error (such as the time term	
this is no error	Ⓟolo Řecent). ︿	

Finally, use an end mark to show where to stop.

the end   ⟨30⟩

---

proofreading

Galley or page proofs usually have little space between lines, so almost all changes must be made in the margin. The point of change is marked by a caret (sometimes by a check mark), with the correction or correction symbol in the margin.

When a line requires more than one change, marginal corrections are assembled, singly, in the proper order with slash marks separating each one from the next. However, if the number or juxtaposition of changes seems likely to be confusing, the best course is to kill the entire word or line and insert the proper form in the margin.

meaning	marginal symbol	mark in type
insert comma	⌃	Standard symbols⌃ used wisely⌃aid the
insert semicolon	⌃	typesetter⌃others too. A good rule to re-
insert colon	⌃	member⌃type is set letter by letter and
insert quotation marks	⌄⌄	⌃space by space⌃and so each letter and
insert apostrophe	⌄	space must be absolutely clear. It won⌃t do to be ambiguous.
insert hyphen	=	Well⌃edited copy is marked to show
insert 1-em dash	⁄m	the length of each dash⌃ as in the life
insert 1-en dash	⁄n	span of the printer Gutenberg, 1400—⌃ 1468.
insert question mark	?	Doesn't that approach seem logical⌃
insert exclamation point	!	Of course⌃
insert parentheses	( )	Parentheses⌃as in this example⌃are
insert brackets	[ ]	often needed; brackets⌃editor's note⌃are
insert braces	{ }	less common. Braces⌃like this⌃are rarer still.
delete, reduce space	⌔	Editors must mark deletions⌿ pre-
delete, close up entirely	⌔	cisely, bearing in mind⌿xx that some⌿one must decide what to do with
reduce spaces	⌒	any ⌃ space ⌃ left ⌃ over.
equalize spaces	= #	The⌃spacing ⌃ can be ⌃ wrong ⌃ in many ways an-

46

Instruction	Mark
add space between words	#
add space between lines	#
restore deleted matter	be
use normal paragraph indent	¶
do not indent	
do not indent	(flush)
no new paragraph; run in	(no ¶)
start new line	(run over)
run over to the next line	
run back to previous line	(run back)
indent 1 em	⊐
indent 2 ems	⊐⊐
set flush left	⊏
set flush right	⊐
center	⊐⊏
straighten type	=
push up	⊓
push down	⊔
align type	‖
turn to proper orientation	⊙
with ragged left margin	}
transpose to order marked	(tr)
transpose words	
transpose letters	(tr)
author, please clarify	(auth?)
missing copy	(copy out)
use figure, not word, and vice versa	
spell out (or abbreviate)	
set in 9-point Optima on an 11-point slug, 26 picas wide	9/11 opt x26

also the instructions by the editor must be precise.

Spacing between lines must be changed sometimes, and often a deleted word must restored.

Indicate paragraph indentation where necessary.

Or show that you don't want the normal indentation.

If there can be doubt, mark "flush". Or "run in" if appropriate.

You can do this: start a new line.

You can run a letter or letters over to the next line, or back to the preceeding line.

Indentations may be marked like this and this.

This way for flush left, ⊏ this for flush right, and this for center. ⊏

Sometimes type is set unevenly and then lines must be reset to straighten the letters or words.

Crooked margins must be aligned in page proofs (it doesn't usually matter in galley proofs). And sometimes a letter is set upside down.

Ragged margins may be set this way for effect, with the wavy line properly marked for instructions.

Sometimes words and letters are transposed, but marking clear will aid correction.

If only the author can answer a question—as of the spelling of a personal name like Psmith—a marginal note will direct his attention to the point.

If the typesetter has skipped several sentences or a longer part, to do is to refer him to the original copy.

It is customary to circle numbers and words, as in 7 and seven and in mm and millimeters if you want the typesetter to "do the opposite" but be careful, for circling also means "do not set" as in the next paragraph.

Marking type specifications calls for a special language to make possible both brevity and accuracy. No publication uses all typefaces and sizes, but the most common form of a given face is called

change to roman type	(rom)
set in bold face	(bf)
set in italic	(ital)
set in small caps	(sc)
set in caps and small caps	(c&sc)
set in bold caps	(BC)
change to lower case	(lc)
change to upper case	(UC)
change to proper face and size	(wf) ⊘
reset broken type	
set superscript and subscript	
that is not an error	(foto)
this is the end	(30)

(sometimes inaccurately) roman.

Bold face is often used for sideheads and the like. Genera and species names are printed in italic, as in Atrypa reticularis.

Other special uses call for small capitals and caps and small caps, and for bold caps.

Capitals can be changed to lower case, and lower case to capitals.

A type character of the wrong face or size can be marked 'wrong font'. Broken or defective type should be pointed out, too.

Subscripts and superscripts, also called inferior and superior figures, occur in such things as . And anything the typesetter might take to be an error, as the time term Recent, must be marked.

Finally, it helps the typesetter to use an end mark to tell him when he's finished the job.

(30)

---

proofreading

The best tool for an accurate proofreader is a suspicious mind. Mistakes cost money if they are repaired, and embarrassment if they are not. One earth-science journal has a "dollar deductible" cost policy: all author's changes beyond $1 per page are paid by you, the author. One commercial publishing house is reputed to give its editors five free errors per book. If you let a misspelled word remain in the manuscript, or missed a style mark, or change your mind about the syntax, you can do it five times free. After that, you, the editor, pay. At $10 an error, the house has careful editors.

Whether or not the author or editor has to pay, somebody does; and it behooves us all to check and recheck. Editor and author must each assume that no one else will read the proof; that he alone is responsible. He should read proof as if he were petting a porcupine: very, very carefully.

Special headings, tables, and the like may be set on Monotype machines. At one time "monotype" meant that the typesetter picked the letters out by hand, one by one, to set in place. Today, in large shops, copy is punched on a tape, which instructs an automatic type-selector to set the type. Even so, setting in Monotype is slow and expensive.

hot type

Proofs from hot-metal machines that are sent to you for correction are on long "galley sheets", usually printed on newsprint directly from the Linotype or Monotype metal. You must mark the proof clearly, both in the body and in the margin. Make your corrections neat but conspicuous. (If the printer doesn't notice them, he won't make them.) But remember: any change in a Linotype line means that the whole line must be reset. If it does not come out evenly—if it will not justify—then additional lines will have to be reset.

cold type　　There are many brands of cold-type composing machines, but there are three main types: (1) direct image, in which only one copy is produced, and corrections must be made by razorblade and glue; (2) photo-image type, which produces only one copy; and (3) electronic methods, by which copy-producing tapes can be stored and reproduced on request.

All three kinds of machines produce copy ready to be pasted down for printing by offset methods. Copy from electronic machines can be corrected by reinstructing the storage tape and "replaying" the entire copy. Corrections for photo and direct image machines go back to the operator, who either resets the lines required or produces words or letters to be used for correction. It is a delicate job to make such corrections and, although they may cost no more than corrections in hot-type copy, the burden of making them may fall on the author or editor. For this reason, cold-type corrections seem to be much more trouble.

Headings for cold-type composition may be hand- or machine-lettered, derived from photographic image, or made up letter by letter from "rub-off" or waxed sheets.

All these cold-composition proofs should be handled with care; they constitute the exact copy that will be photographed to print the book, and each thumbprint, pencil mark, and coffee drip will show.

No matter how the manuscript is to be printed, authors and editors should send in as clear, unequivocal copy as possible, and should practice self-control in making "nice" but unnecessary changes.

the problem of hyphens　　In the near future, there may be no typesetter working on your typescripts. Automatic typesetting methods, which read from typed copy, are in use throughout the country, and are being refined rapidly. The main difficulties now being overcome relate to problems of hyphenation, spacing, and correction.

Some machines can carry rules for hyphenation in their memory banks; even so, they can make mistakes (such as between pro-duce and prod-uce). Mistakes are costly and troublesome, and require a living person to deal with them.

If you don't mind ignoring tradition, it is possible to solve hyphen problems by simply doing without hyphens. But if you want to hold to the traditional method of hyphenation—which theoretically is a service to the reader—you must be aware of the problems that arise from splitting a word at the end of a line.

In the days when only printers printed, the problem was not great, because they knew, as a part of their business, how to split words properly. Today, each writer may be his own typesetter (as when typed copy is to be photographed for abstracts volumes) whether he knows the rules or not.

Basically, the principles are these: split words by syllables (consult the dictionary when in doubt); try to avoid splits if at all possible; if not, try to distribute the splits over the page so that they do not call attention to themselves (no two following one another, for example); make it as easy for the reader as you can, avoiding word splits at the ends of pages, or before tables or illustrations; break words in the most logical place you can, so that the reader will not guess the finish incorrectly.

In geological work, you should mark the typescript and watch the proof to make sure the typesetter has not improperly run together rock and mineral terms or other technical phrases. If, in the typescript, the first half of such a term as dihexagonal-dipyramidal appears at the end of a line, it may very well show up as dihexagonaldipyramidal.

permissions Marking copy and checking proof are not the only hurdles for the author and editor. If your typescript contains quotations of more than a few words, or has illustrations taken from or based on ones that have been published previously, someone must see that permission of the copyright owners is given for reprinting. If you, the author, have obtained the written permission of the copyright owner (generally this is the publisher—the Paleontological Society, for example), well and good; you should send a copy of the permission to the editor. If you have not, then the editor must undertake to get it. In any event, there must be a written letter of permission on file before you publish anyone else's words or drawings.

It sounds rigid, but it is the law—and the golden rule.

copyright The basic idea of copyright is simple: copyright is an aid to legal enforcement of fairness. Strictly speaking, you do not "copyright" your book: you do print the copyright notice in a conspicuous place in the front of the book (often on the back of the title page), obtain the proper form from the Register of Copyrights, Copyright Office at the Library of Congress, Washington, D.C. 20559; file the form with the specified number of copies of the finished work and the fee, and wait for the form to come back.

The notice should take this form: Copyright © 1973 American Geological Institute. (The © is an international symbol.) And it *must* appear in *all* copies that are distributed in any manner—all copies, including such things as mimeographed versions of a text being tested in the classroom. If any do not carry the notice, you may lose copyright. (For precise rules, write to the Copyright Office for its booklet *General information on copyright*.)

LC number Often associated with the copyright notice in books are the Library of Congress (LC) catalog number and the International Standard Book Number (ISBN). To obtain an LC number, write to the chief of the Card Division, Library of Congress (Building 159, Navy Yard Annex, Washington, D.C., 20541) at least a month before publication; give the exact title, approximate number of pages, date of publication, and a short description of the contents.

ISBN For an ISBN, write to R.R. Bowker Company (1180 Sixth Avenue, New York, 10036), which will (as an example) issue a unique block of 100 numbers of which 0-913312-01-2 might be one. In that example, the 0 designates the English-speaking world, the 913312 the publisher (in this case the American Geological Institute), the 01 the "title identifier" (from the unique block mentioned before), and the final 2 a "check digit" for computer verification of the logic of previous digits. Later, you report the book title and number to Bowker, which acts as the repository for the United States and Canada and replies to inquiries from, say, Uganda. You may also want to use the ISBN as your stock number.

CODEN A systematic standard of unique 5-letter designators for the world's serial journals is called CODEN. The system is maintained by the American Society for Testing & Materials (at 1916 Race Street, Philadelphia, Pa., 19103), which publishes and disseminates lists of existing CODENs with the journal titles, *in extenso*, and upon request (for a small fee) will issue a new CODEN for any periodical not yet designated. A new CODEN thus assigned becomes part of the standard listings. Some users who depend on computer storage and manipulation add a sixth, or check character, which is determined by a pre-set numerical formula that assures computer-rejection if the CODEN has been erroneously keyboarded. Unfortunately the check character has not been standardized.

ISSN　　A system of International Standard Serial Numbers (ISSN) is being developed in Paris with worldwide national centers. Although the new system is being supported by Unesco and many publications carry ISSNs, it seems years away from practical application.

colophon　　Increasingly common on the copyright page of books is the colophon: a concise list of such information as the name of the designer, production manager, typesetter, printer and binder, the name and characteristics of the main text typeface and its size and leading, the method of composition, the name and weight of the paper, and the press run. The colophon is more than a courtesy; its information is extremely useful when reprinting or a new edition is proposed, or when a similar book is being planned.

design with type

Use of type in design is both an art and a science. A logical approach is to treat typography first as a science and learn the nature of the thing and the mechanical rules that control its use. Out of that should come an understanding of the grey area between the black and white of the "rules", for somewhere in that area the science becomes a craft and then an art.

First, some basic information. Printing measurements are based on the "point", which in the English-speaking world is 0.0138 inches or 0.351 mm. (In most of Europe, a point equals 0.376 mm.) For almost all work, type sizes are given in points, and most other measurements are given in 12-point units called picas. The pica is measured both horizontally and vertically on the page, but the point, which is used for line spacing and type size, is measured only vertically—never for horizontal space.

- *1 inch = 6 picas = 72 points.*
- *Type-specimen books give the average number of characters per pica for many typefaces.*
Those facts are the basis of all copyfitting and much design with type.

Usually it may be assumed that 1 inch = 6 picas (72 points), but if you look at the right end of an ordinary pica rule (inches on one edge, picas on the other) you will see that 72 picas is slightly less than 12 inches—about 1/32 inch less. That discrepancy has practical importance, for paper-making machines and presses are designed in inches. As a result, the outside dimension of any piece of printing design is usually given in inches, but inside dimensions, being based on type sizes, are given in points and picas.

An example: the trim size of this book was specified as 7 1/8 × 8 1/2 inches (here we follow the mathematical practice of giving the horizontal dimension first), but the type is measured in points, in this case 11 points per slug. (When cold type is used, consider the slug the distance from the bottom of the descenders of one line to the bottom of the descenders of the next line above. Because type is measured in points, the type page was specified in picas—37 1/2 × 46. A 7 3/4-inch column would accommodate 51 lines, but slightly less than 1/16 inch is left over, and the difference must be accommodated either by shortening the column or—in hot type—by inserting thin strips of type metal that are called "leads".

The amount of leading—that is, the size of a slug in addition to the type size—determines the distance between lines. It does more than that. It adds to the legibility by helping the reader's eye stay on the line, although too much leading wastes space and produces the impression that the type is for a child's eye. Leading also contributes to type color or texture, a characteristic you can see best by placing several publications on a table and standing just far enough from them that you cannot read the text.

Now, before going on to typefaces themselves: here is danger. The best design is done by exploiting the possibilities of a limited number of faces; however, almost any type-specimen book has a large and attractive variety of faces, and the variety often intoxicates a beginner, who thus spends too much money on too many faces and produces too much clutter on the page. Simplicity is best.

Optima

In this manual the name of the text face is Optima; the size used is 9 points high; each line of type is centered on a slug 11 points high; the measure (length of slug, or line) is 26 picas. Thus the typescript was marked "9/11 Optima × 26″ (read as "9 on 11 Optima by 26 picas").

Times Roman

A typeface more familiar to many readers is Times Roman (originally designed for the *Times of London*). It belongs to what is known as the Roman class of type; it has serifs—the thin strokes at the ends of the main components of the letters—and graduated thickness of strokes. The Roman class is based on writing by carving letters in stone by a square-pointed tool.

Techno; Stymie

Lorraine Script

BROADWAY

Caslon

Baskerville

Bodoni Bold

g a e p t
g a e p t

**Univers Extrabold
Extended**
Univers Light Condensed
Univers Medium Extracondensed
Univers Light
Univers Bold
TIMES SMALL CAPS
Times italic

bp
bp

x

x

x

copyfitting

Other major classes:
• Abstract, with all lines of the same thickness, including block or square serifs, if any. Faces without serifs are referred to as sans serif.
• Cursive, based on slanting writing and sometimes restricted to letters joined in a continuous line.
• Decorative, in which elements are exaggerated or unexpected features are added.

The various Roman types are distinguished by the character of their serifs and by the relative weight of the thick strokes and the thin. Names given the subdivisions of Roman sound as if they indicate relative age of the design but usually they do not. One common classification:
• Old Style, with a flowing transition from thick strokes to thin strokes, and bracketed serifs.
• Transitional, with greater contrast of stroke thicknesses, and less-pronounced serifs.
• Modern (including many designed long ago) with great contrast between thicks and thins, and geometrically precise serifs with no brackets.

Often the classes and subdivisions are imprecise, and serve only to indicate general characteristics. The differences between individual faces may be quite small, but it is worth noting that the lower case g is often the most distinctive character. Other letters likely to be distinctive are a, e, p, and t.

Many variations are to be found within the design of a single face, such as extended, normal, condensed and extra-condensed (based on relative width of the letters); lightface, standard, bold face, and extra-bold (based on weight of strokes); small capitals and italic. Slanted letters are called italic or slant; vertical letters are called roman (but note the potential for confusion with Roman). These variations also contribute to the type "color".

Text type, or body type, refers to the size used for the main part of a given publication; generally it consists of type up to 18 points. Display type refers to larger sizes, used for chapter headings, advertising headlines, and the like.

The point size of a typeface is less definitive than you might logically assume. In theory the size is the distance from the top of an ascender letter such as b to the bottom of a descender letter such as p—but in one face the "bowls" of the b and p may be much larger than the bowls of an otherwise similar face.

A more meaningful measurement for some purposes is the height of the lower-case x, a measurement rarely given in type-specimen books. However, a letter in 10-point Times Roman can be counted on to be slightly wider than in 10-point Garamond, as shown in the comparative lengths of the lower-case alphabets. This relative difference is usually expressed in terms of the average number of characters per pica (c.p.p.) for average text matter. For 9-point Optima, the c.p.p. is 2.80; for 10-point Optima, it is 2.55. Many type-specimen books carry the c.p.p. for each font—a great help in copyfitting, or calculating how much space will be filled after the type is set. (A font is a complete set of characters for one size of one typeface.)

The c.p.p. may be used this way: A typescript is found to consist of 250 lines, each 50 characters long (take care at this point to find the average line length, for it is the source of greatest uncertainty in this procedure). That means the typescript consists of $250 \times 50 = 12,500$ characters; if it is to be set in 10-point Times Roman you divide 12,500 by 2.65 (the c.p.p.) and find that if the type were set in one line it would be 4,717 picas long.

If you plan to use 13-pica lines, you will have 363 lines (4,717 divided by 13); if you use 11-point slugs and 10-inch columns you will get 65 lines per column; therefore your copy, set in type, will fill 5.6 columns (363 divided by 65).

A simpler method consists of merely multiplying the c.p.p. (2.65 in this case) by the proposed line length in picas (13), typing each line to that number of characters (35), counting the lines of typescript, and dividing by the number of slugs per column. This second method does not allow easy adjustment if the copy turns out to be too long or too short for the space available, but you can still resort to the first method.

adjusting to fit   If the copyfitting indicates that the type will run too long for the space, you can • reduce (or eliminate) the leading, • increase the column width, • reduce the type size, • change typeface, • or use some combination of those. Whatever you do, choose the method that will have the least adverse effect on legibility.

Assuming that you are involved in design, it will help greatly at this point to assemble a check list of all elements likely to appear in the printed piece, such as text, headline or title, by-line, identification block, abstract, references, footnotes, credit block, lines drawings, halftones, cutlines, tables, graphs. Calculate the length of each element in column inches, convert the total to columns or pages, and round the total off to the next full page. Then you can make "thumbnails" or preliminary sketches of page layouts and adjust the elements (perhaps changing the size of some) and the remaining white space to yield the "best" (by your standards) arrangement. And be sure to design facing pages *together*, as a single unit.

With a sound knowledge of the mechanics of typography and a little imagination, the design potential is endless.

from ink to paper

"Please come to the front of the room" the journalism professor said to the coed. "Now, freshen your lipstick, and kiss me on the cheek." Then: "Here, class, is an example of letterpress—printing from a raised inked surface."

Turning again to the coed, he said, "You have a Kleenex? Good. Carefully, now—blot the lipstick from my face.

"Here, class," he said, holding up the Kleenex, "is an example of offset printing, by transfer of ink from a flat surface."

Of course the professor oversimplified the distinctions between the two leading methods of printing (the third major method, gravure, is economical, only at much higher press runs than is common in science printing). However, it was a starting point, and in a minute or two his class learned more about the subject than most writers ever learn.

Writers and editors must understand something of the mechanics of the various ways of putting ink on paper. That understanding helps them choose the optimum combination of methods, equipment, cost, and timing. It also helps prevent trouble for others.

Typesetting methods are often classed as either "hot" or "cold"; briefly, "hot type" refers to processes involving molten type metal, and "cold type" to all other methods.

Most hot type is set by molding (or casting)—by forcing molten type metal into a mold called a matrix and then extracting the cooled metal character. This may be done so that an entire line of type consists of a single piece of metal, or so that each character in the line is a separate piece.

Linotype Type for most newpapers and magazines is set by Linotype or Intertype, which are similar machines that are keyboarded much like a typewriter. As the operator depresses the keys, the matrices—usually called mats—fall from overhead magazines and are automatically assembled into a line, with spaces between words being produced by wedge-like space bands forced up from below the line. Hot metal is forced against the row of mats; when it cools, the finished slug falls into a rack, and the mats are automatically distributed to their proper places in the magazine for reuse.

Many books and technical journals are set in hot type by a more complicated machine, the Monotype. This is actually two machines, a keyboard and a caster. On the keyboard machine, the operator produces a spool of punched tape that later controls the caster in producing lines of separate characters.

Monotype The advantages of the Monotype lie in its flexibility. It has perhaps the greatest range of typefaces in the world. Any individual character can be changed, whereas for Linotype and Intertype work any correction requires the resetting of an entire line. The Monotype tape can be stored for use in producing later editions without repeated keyboarding. It can produce a wide range of type sizes and line lengths.

However: the simpler linecasting machines are adequate for many printing jobs. Pay for only the capability you need.

A comparatively primitive device is the Ludlow typecasting machine in which the type is assembled by hand and cast as a single slug in the machine. The Ludlow is extremely common and is much used for large type (display type) as in newspaper headlines.

"Cold type" is any type not involving hot metal; it may even include handwriting. Another form is typewriting (this is also called "strike-on").

The route from the writer's mind to the reader's involves these mechanical processes:

- *typesetting*
- *platemaking*
- *printing*
- *folding and binding*
- *packaging*
- *mailing.*

Typewriter-like machines such as the Flexo-writer and the Justowriter are capable of producing justified lines—all lines are the same length, in contrast with the ragged right margin of ordinary typed material. However, typewriter type is somewhat inferior in appearance and the number of typefaces available is extremely limited. Also, few type sizes are available.

At least one typewriter-computer system produces type that is satisfactory for many jobs with automatic justification, but it, too, lacks the capacity for display-size type.

photocomposition

With the advent of computer technology, many companies have developed machines for high-speed photocomposition. (The link with computers is the huge memory bank needed for hyphening. Human intervention is possible when the machine can not decide where to break a word, but the human element is inadmissible in a system that may set 10,000 characters a second). Perhaps the most widely known photocomposition machine is the Photon, which scans a punched paper tape produced by a keyboard operator (or several operators, in order to keep the Photon busy) and flashes a stroboscopic light through the proper place in a spinning disc—the disc being a master negative bearing 16 alphabets of 90 characters each. The image is recorded on film and a contact photographic print is produced ready for making a printing plate. The Photon can mix 64 fonts and can set as many as 1,000 characters a second.

A more advanced system called the CRT (for cathode-ray tube) is coming into use; in this system, a computer program is written for each typeface and the machine generates each character separately—up to 10,000 a second.

Photocomposition is rapidly displacing "conventional" methods of typesetting. It is making most headway in books, where large quantities of type must be set, in dictionaries and directories, where computer manipulation of the material is required, and in subscription labels, where high speed is important. It is least efficient in small jobs, or where appreciable time must be spent changing fonts, column widths, and the like, as in display advertising.

Proofs and proofreading, which are treated elsewhere in this book, are affected by the different systems of setting type. After linecasters produce type, proofs are pulled directly from the metal. In the case of Monotype, individual characters may be changed. For Linotype, changing a single character involves the complete line. For cold-type machines, from strike-on methods through CRT, proofs usually take the form of a Xerox copy. Any change in the original means that someone must carefully cut out a part of the photo copy and strip in the correction. In short, author's alterations in cold type weigh heavily against the high-speed systems and greatly increase the cost and the time—and the need for careful copyediting.

Note that photocomposition can produce completed pages. However, that is ordinarily suitable only for books, and then only if there has been long coöperation and careful work by the editor, designer, and programmer. No matter how the type is set, the process usually follows these stages:

production route

1. Editor sends marked copy to the typesetter.
2. Typesetter sets the type and pulls galley proofs (usually about 18 inches of type, regardless of editorial divisions).
3. Editor (and sometimes author) reads and corrects the proofs.

4. Editor cuts up the galleys and (allowing for any changes in length due to changes or corrections) pastes them up on artboard, producing a detailed dummy that shows exactly where each line of type and each illustration is to appear.

5. Using the dummy as a guide, the printer (if hot type is used) assembles the type into page forms and pulls page proofs. If it is cold type, an artist cuts up the photo copy and glues it accurately in place.

6. The editor checks the page proofs, or electrostatic copies or photocopy pages, or the actual pages, or a blueprint or similar copy.

7. The printer makes plates and prints the job.

platemaking

Although the original hot type may be inked for printing, temporary (remeltable) type, such as that produced by a Linotype machine, is commonly used instead. Because offset printing is rapidly replacing letterpress, for most hot type and all cold type, plates are made for the actual transfer of ink to paper.

A great many kinds of plates exist; but there are two basic categories: duplicate plates (or stereotypes) and photomechanical plates.

Duplicate plates are widely used by newspapers that are printed by letterpress. The type is locked in a page form, or chase, and a heavy sheet of special paper is forced against it in a press. The resulting paper mold is used to cast a type-metal plate (usually several, so that several press units can print the same thing at the same time) for printing. These plates are often cast in semicylindrical molds in order to adapt them to cylinder presses.

Another way to make a duplicate plate is to make a plastic or sheet-lead mold of the original type, spray it with silver nitrate, and place it in a chemical bath that will electroplate it with copper. Backing is added to stiffen the copper for the stresses of printing.

Photomechanical plates are used primarily to reproduce artwork. The simplest kind of artwork is a line drawing, which is anything consisting solely of black and white with no intermediate shadings of grey. To make a photomechanical plate from line art, simply use a film negative of the line copy (which may include type), then project the image onto a metal plate with a photosensitive coating. Finally, treat the plate with chemicals that etch away the metal in the areas that are not to print. Such plates are known as line cuts.

screens

Halftone cuts are made from copy characterized by continuous tone, such as black-and-white photograph, or a wash drawing—anything in which there are grey tones, intermediate between black and white. Such copy is photographed through a glass plate carrying a screen of very fine lines, usually crossing at right angles, which breaks up the image into tiny squares.

The squares in black areas are full size, in white areas almost nonexistent, and in grey areas the size depends on the depth of color.

Halftone plates are exposed and etched in much the same way as line cuts, but the result differs in that all areas to print grey or black will be covered with a pattern of squares that, to the unaided eye, resemble dots: the darker the area, the larger the dots (up to the limit of the screen size). Only the dots stand high enough to touch the paper.

The size of screen chosen depends on the paper to be printed: most newsprint requires a coarse screen (55 to 85 lines per inch) to keep the space between the dots from being plugged with ink. Smoother papers may permit a 150-line screen or more, resulting in finer detail.

Note that a printed halftone usually cannot be "picked up"—rephotographed—because when a new plate is made the original screen will in effect be rescreened; the result is loss of resolution; and the rescreening may produce noticeable moiré effect. Nor can the original screened negative be enlarged or reduced more than slightly. Normally, line work and halftone work should not be photographed together on the same negative, as screening breaks up lines into fuzzy rows of dots. They should be photographed separately and then reassembled.

letterpress　　Printing by letterpress usually means starting with hot type, but cold type can be considered line copy, and so photomechanical plates can be made from it (even though that is seldom economical).

For offset lithography, too, the starting point can be either hot type or cold. Cold type is photographed directly. Hot type can be treated in many ways, but the aim is to obtain an image that can be photographed—in a sense, to convert it to cold type. Perhaps the most common method is to pull a reproduction proof (called a "repro") on a smooth paper. In the Scotchprint process, the proof is pulled on a plastic sheet. In the Brightype process, the type is blackened, the printing surfaces—only—polished, and the type is photographed directly. Variations abound.

As in photoengraving for letterpress work, the printing image is projected onto a metal plate with a photosensitive coating. But there is an important difference: for a letterpress plate, the areas to be printed are etched away until only the printing areas will touch the paper when on the press. For a lithographic plate, the chemical treatment differentiates the plate into oil-repellent areas (non-printing) and water-repellent areas (printing); the ink used for printing has an oil base, and the oily ink and water—which do not mix—are applied together.

offset　　That much is lithography. On most presses using the offset principle, the plate is wrapped around the impression cylinder, which transfers the image to a rubber blanket; the blanket, in turn, transfers—offsets—the image to the paper. Hence the common shortened form "offset".

Much in the same way that photocomposition is becoming dominant in typesetting, offset is taking over in printing. Although geologists were once cautioned to avoid using artwork because of the expense of making zinc plates, the comparatively low-cost plates for offset are making photos and other artwork much more common in the science press.

ink　　Now, briefly, consider ink: no one kind will suffice for all kinds of work; ink for letterpress is quite different from ink for offset, but, luckily, ink is less a problem for the editor than for the printer. (It is safer to refer to "inks" than to "colors", as some printers count black as a color and others do not.) However, it is worth knowing that the choice of inks is by no means limited to the standard colors shown in manufacturers' catalogs; almost any color can be matched and custom blended at a cost not prohibitively greater than black.

imposition　　The question of inks comes up often when black and one or more additional inks are to be printed on the same page. Extra inks can open up an enormous range of possibilities, but in order to cut costs and simplify matters it must be borne in mind that each ink requires a separate plate. If the number of pages is so large that more than one plate is needed to print only one ink, it may be possible to arrange the layout of pages (the imposition scheme) so that fewer plates are needed for the second ink. But that only points the way to the many uses of a complex and specialized subject.

Imposition consists of assembling and arranging all the pages that are to be printed on one sheet of paper. It must be done in such a way that after the sheet is printed and folded each page will appear in its proper position. (The pages that are folded from a single sheet are collectively called a signature.)

As a few minutes with pencil and paper will show, even a booklet of only a few pages can be arranged in several imposition schemes, and the number of possible schemes goes up sharply as the number of pages increases. The one to use depends on factors such as the press and folding machine to be used, the number of pages on the sheet, the desired arrangement of signatures, and the characteristics of the paper.

Editor, designer, printer, and binder must all agree on the best imposition scheme for the job. Efficiency of binding is often the most important factor, but the designer may have an overriding need for, say, a second ink on certain pages—perhaps a map that requires color for clarity—at the lowest possible cost.

For example, one imposition scheme may allow a second ink (using a given number of plates) on only consecutive pages in the middle of a signature. Another scheme may allow the second ink on the pages near the beginning and the end. And so on.

If a second ink is required on certain pages of a publication, their position will suggest an imposition scheme. That scheme will in most cases allow use of the same ink on certain other pages at little or no extra cost.

paper
Quite possibly the most complex single factor in printing is paper. The editor must learn what he can, but he should call on an expert to make most choices of paper, or at least to narrow the choices.

Briefly, paper is sold by weight, and weight is generally given as the weight per ream (500 sheets). But different papers come in different sizes, such as 25×38 inches for book papers and 20×26 for cover papers. (Unfortunately, U.S. manufacturers do not yet follow the British practice of specifying grams per square meter regardless of sheet size.)

Weight and cost aside, the best paper depends largely on the printing process. For example, many old and still popular typefaces yield their best appearance when printed by letterpress on a soft paper. Finely screened halftones may lose detail if printed on any other than a smooth, hard finish, but the same paper may produce too much glare for text.

A short list of common papers may suggest other factors that should be considered in making a choice for a given publication: antique (rough surface, for bookwork), machine finished (polished surface, for halftones up to about 100 screen), low sulfur (for long-lasting books), surface-sized (to prevent absorption of water in offset printing), coated (usually with kaolin, for great smoothness).

Generally, any paper can be printed by letterpress, but surface sizing on lithographic papers may cause trouble for letterpress work.

binding
Folding and binding, falling as they do near the last stages of print production, are often unduly neglected. Much trouble can be saved merely by providing the printer—early in the game—a blank dummy of the finished work, folded and trimmed to the exact size visualized.

The blank dummy can also be used to check envelope sizes—it should fit a standard envelope or box, or special ones must be made. Check and decide at this point rather than after the job is printed, bound, and delivered.

Separate maps, included in the back, or issued in an accompanying container, may be folded by the printer in such a way that they are difficult to use and nearly impossible to refold. Give the printer a sample of the way you want the maps folded. Neat accordion-folds not only make the map a great deal easier to use but also increase its life expectancy.

Binding methods include saddle stitching (usually by wire staples through the centerfold, side stitching (as by an ordinary office stapler), section-sewn or Smythe binding (folded sections are saddle sewn, with thread, and individual sections are then sewn together—most casebound books are bound this way), and perfect or threadless binding (all the folded sections, or signatures, are gathered, the back fold is trimmed square, and the individual sheets are glued by their back edges to the cover—see the binding of this book).

Caveat: printing technology, which changed only gradually in the first 500 years after Gutenberg, has in the last decade or so been developing at bewildering speed. No account of printing methods can be up to date for long.

writing reviews

Reviews of books for the general public are written by professionals who live by their trade. Those people cover creative works, such as novels and poetry. In all these cases, the reviews reflect the taste or intellectual and social judgment of an individual who knows (even when he is over his depth about the content) that his reader wants his personal reaction as a helpful guide to his choice of reading. Such reviewing is not our model here.

We speak of reviewing scholarly books as a means of education in the broad and disciplined sense. These are books that purport to add something to the corpus of man's knowledge, books written for serious students and scholars, books that purport to be original contributions in analysis, or in form, emphasis, and content. A quite different responsibility now rests on the reviewer.

Here the reviewer is not a professional writer; rather, he is a specialist in a certain field of knowledge. He is chosen for that knowledge, for his critical awareness of the work and thinking going on in the subject, and for his sensitivity to advances. This specialist may be inexperienced at review writing; indeed the book-review sections of our journals reflect this fact.

Where does the specialist reviewer start? How may he best proceed? What are his responsibilities?

He should remember, as his first commandment, that he is responsible to his readers, who are in general fellow professionals, students in the broadest sense. He is not responsible to the author, who having published must be prepared to accept honest, objective, competent evaluation of his work. Nor is the reviewer responsible either to the editor, board, or group that chose him, or to himself (except for doing a competent job). Unlike the general book reviewer, he is required to subordinate his personal feelings to objective appraisal. A subjective opinion alone is not enough.

What should this kind of review tell its reader? Simply what he needs to know. The vital statistics at the start—title, author, pages, date, the publisher's name and address, and the price—are obvious. What else?

Obviously, the reviewer should read the preface and any introductory statement, to be aware of the author's intentions as to reader and purpose. Otherwise, the reviewer is apt to play king, and say how he would have written a book on the subject, which is not what the reader of a *review* needs.

First, what reader is addressed? Patently, *An introduction to college geology* as a title speaks for itself. However, *Cranial muscles of* Lucanus cervus *in Iceland* will require some elucidation. The review reader needs to know the scope of the book and its level, and if the beginning-geology book presupposes calculus and physics, or comparative anatomy, those facts are relevant to note in the review, for they restrict the readership.

Second, given the book's purpose of addressing itself to a certain topic *and* a certain reader, what original value has its content? What new knowledge or what new emphasis or insight does the work offer? The singular nature and *raison d'être* of any book is that it is—or should be—singular. In what respects and to what degree is the book unique and its originality of value? Comparisons with other published works may be helpful.

Third, how successfully, given the reader, subject, and singularity, does the author execute his purposes? Does he offer his information or analysis logically, lucidly, and consistently? If weak in any such respects, is the book still worth the effort of study?

About this matter of lucidity: as emphasis on rapid and frequent publi-

A review should include these facts:
- *exact title (from title page only)*
- *full name of author or editor*
- *publisher's name and address*
- *year of publication*
- *edition*
- *number of pages*
- *page size if unusual*
- *number of illustrations if appropriate*
- *type of binding if significant*
- *price*

and an appraisal of:
- *intended readership*
- *scope and level*
- *strengths and weaknesses*
- *originality and lucidity*
- *singular features*
- *execution*
- *comparability with similar works*
- *accuracy*

and
- *reviewer's name and address.*

cation has grown, many authors have slipped into bad habits. There has been haste to grind out writing with no respect for lucidity of organization and exposition. One could describe many of the results as written in academese. Technical language that spells out meaning precisely, yes; technical Pelion upon Ossa, no. Lucidity is still the *sine qua non* of a good book.

About accuracy—and errors, large and small: small errors, if prevalent, call for comment. Yet, since there are statistically millions of possibilities per page for small errors, so that *no* book is ever free of them, these are not usually worth much comment. It is the validity of the content in general—the large matters—that the reviewer must evaluate.

Given the achievement of those tasks, a reviewer has the right and duty to express—parenthetically, as it were—his own personal reactions to the work. Yet here there is a fine line he must observe. He must be objective. He may show favor or regret, enthusiasm or distress. But he should, in any event, be reasonable, not subjectively personal, always remembering that his primary responsibility is to his readers.

The specialist reviewer is often chosen by some board or editor, with the idea that the man's work and the book's content have a professional affinity. The person chosen usually feels a duty to carry through. But at times, to act on this feeling is not valid. If a first reading indicates to him that he is not suited to the task, he will best serve his fellows and himself by withdrawing. Otherwise he may write, as has been done, that the book is difficult to review because it is so "different", a departure from what has been done. Any good book should be a departure.

Specialty reviewing is an underdeveloped art because of the responsibilities of the reviewer and the fact that he is not chosen for his talent as a writer. Yet he is an educated person, and as such he should do for other specialists as he would have them do for him when their turns come to review his books.

Of the making of good books and bad, there seems to be no end. Any review, favorable or not, is publicity, and results in sales. Good reviews may do more than help a reader choose before he reads. When a reviewer helps his readers to eschew the bad books, to be aware of the reasonably good ones, and to "discover" those that are singularly good, he is also helping to achieve higher standards and (let's hope) the making of fewer and better books. The practice of this underdeveloped art may thus become a rewarding responsibility.

writing for the press

- *To persuade an editor to use your press release, make it easy for him to use.*
- *Write to inform the readers, not to impress the subject of the release (or your boss).*
- *Check, re-check, and check again.*

The first function of a press release is to get in print. The surest way to get it in print is to design it for the editor who will decide whether to use it: the easier it is to get it in type the more receptive he will be.

Mechanical aspects aside: the editor won't use your release unless you write it to inform his readers. "Edward L. James, president of Amalgamated Oil & Gas, announced today the discovery of a new field in Cook Inlet." Few readers care about James or his title, and the news isn't his announcement. Tell the reader that "Amalgamated Oil & Gas has discovered a new field in Cook Inlet." If you really must mention James, do it later.

Get the most important information in the lead. Keep the lead short. Keep paragraphs short (newspapers have narrow columns).

Bear in mind that if the editor doesn't use the entire release he's likely to cut it from the end.

Don't underestimate the editor's or reader's intelligence, or overestimate their information.

Double-check all facts, figures, and spellings. Pay particular attention to any arithmetic (if you say 11 companies are drilling off Java and then list them, count the names). Verify all proper names, for the editor may not be able to check them.

If you include a photo, remember that it may be used with the release; or not at all; or alone. Write cutlines for the photo with that in mind. Give the news but do not point out the obvious ("shown shaking hands") or attribute a fictitious activity ("shown discussing earthquake damage" if the subjects are obviously doing nothing but posing for the photographer).

For any photo, type the date, any necessary credit, and full identification, and tape that information to the back of the photo.

If you plan to send your release to a single publication, study that publication and find how it uses information like yours. Then write your release that way.

In any case, following the suggestions below will increase the chances that your release will go into print instead of on the editor's kill spike.

- Type "press release" at the top of the first page.
- If you want the news held until later, write something like "HOLD FOR RELEASE ON MAY 1, 1974", and put that, too, at the top.
- Leave blank at least 18 picas above your text so the editor can write a headline.
- Leave blank at least 9 picas at the left-hand margin so the editor can write instructions for the typesetter.
- Use a dateline, such as "Menlo Park, Calif.—Oct. 12, 1974", so the editor will be sure when and where the news occurred.
- Double-space all copy so the editor can change style easily and clearly.
- Don't underline anything, for few newspapers have italics.
- If the release is longer than one page, type "more" below the end of the last line on all but the last page.
- Never type on the back of a page.
- Avoid breaking words from line to line, and never break technical words (or you may find an unintended hyphen in print).
- As a convenience to the typesetter, don't break a paragraph at the bottom of a page.
- Use an end mark—"end" or "30".
- After the end mark, close with a source of more information—name, title,

address, and telephone number. (Consider giving both home and business numbers.)

- Staple all pages together (a survey of editors reveals that loose sheets cause more annoyance with press releases than anything else).
- If there is a photo, tell the editor about it after the end mark—not in the text.
- Proofread the final copy. Do it yourself—don't ask the typist to do it. Ask a third person to read it, too. As an extra precaution, read it aloud.
- Finally, ask yourself: Have I written the news for the reader? Have I made it easy for the editor to use?

reference shelf

Any editorial office should have a shelf of reference works on subjects ranging from writing through printing—those in addition to books in the field of the publication's subject matter. This section of *Geowriting* might be considered a shopping list for such a reference shelf; in it we list important works in various categories and compare their merits.

This section is also the equivalent of a selected bibliography for *Geowriting*; as such, it might have been entitled "Where to dig deeper".

style books

One of the main tools of any editor or writer is a style book. Yet, style books are notoriously hard to use, and usually each one is intended for a single purpose and so is unlikely to meet the needs of others. Almost every editor adopts a style book and either adapts it to his needs or tries (wrongly) to force his copy to fit the style book's mold.

One of the most widely used is the *U.S. Government Printing Office sytle manual* (revised January 1973; Washington, D.C., 548 p.). This is a highly useful work, with suggestions to authors, and guides to capitalization, spelling, punctuation and so on, plus guides to the typography of many foreign languages. However, its usefulness is diluted by instructions applicable only to the *Congressional record* and other specific government publications, and by the fact that it tries to be all things to all editors. For most geologists it will be a handy reference but not a bible.

Another widely used guide is the University of Chicago's *Manual of style for authors, editors, and copywriters* (12th edition, 1969; University of Chicago Press; 546 p.). It, too, is used for an entire stable of publications, including the *Journal of geology*. It has given rise to at least one other style book, Kate L. Turabian's *Manual for writers of term papers, theses, and dissertations* (third edition, 1972; University of Chicago Press; 164 p.), which is said to be the most widely used standard in college departments of geology.

The CBE style manual (by the Council of Biology Editors, third edition, 1972; American Institute of Biological Sciences, Washington, D.C.; 297 p.) is not merely a style book but also includes sections on writing, copyfitting, editing, and indexing. It suffers somewhat from trying to be too inclusive, but is a good starting point in any case and especially so for, say, paleontologists.

In geology, the best known style reference is *Suggestions to authors of the reports of the United States Geological Survey* (1958; Government Printing Office, Washington, D.C.; 255 p.). This is the fifth edition of the only general style guide for publications in geology. However, it was written primarily for the purposes of the Survey, and is out of print and out of date. Some users prefer the fourth edition, by B.H. Lane (1935); either may be useful but neither can now be called a standard.

journal style

Many technical journals publish their style rules every year or so in the journals themselves. Among examples are "Suggestions for contributors to the *Journal of sedimentary petrology*" (by Gerald M. Friedman, 1965; *Journal of sedimentary petrology*, v. 35, p. 5–11), and "Note to authors concerning form and procedure for the *Journal of paleontology*" (by Erwin C. Stumm and Robert V. Kesling, 1957; *Journal of paleontology*, v. 31, p. 1,019–1,028).

One reference is prized by editors because it concerns not only style but also grammar, usage, and printing practice; it is *Words into type* (by Marjorie E. Skillin, Robert M. Gay and others; revised edition, 1964; Appleton-Century-Crofts, New York; 596 p.). Unfortunately its coverage of printing technology is out of date.

The New York Times style book for writers and editors (edited by Lewis Jordan, 1962; McGraw-Hill, New York; 124 p.) is a dictionary-style guide; if you want to determine when and how to abbreviate "Missouri" you look up not a section on states or abbreviations but "Missouri". It is easily adapted to nontechnical needs and is a good reference as well.

The guide perhaps most widely used by nontechnical scholarly journals is *The MLA style sheet* (second edition, 1970; Modern Language Association of America, New York; 48 p.), with more than 2,600,000 copies in print.

If your subject is education, a good reference might be the *NEA style manual for writers and editors* (1966 edition edited by Jane Power and others; National Education Association, Washington, D.C.; 76 p.).

Geologists who hope to publish in non-geology journals or who specialize in non-geology fields may find it worthwhile to obtain works such as *Manual for authors of mathematical papers* (1962; American Mathematical Society, *Bulletin of the American Mathematical Society*, v. 68, no. 5). Among others are *Style manual for guidance in the preparation of papers for journals published by the American Institute of Physics* (1962; American Institute of Physics, New York; 42 p.) and *Handbook for authors of papers in the research journals of the American Chemical Society* (1965; American Chemical Society, Washington, D.C.; 93 p.).

technical writing Even though the principles of good writing are largely independent of subject matter, there is a host of books giving instructions for writing about technical subjects or for technical journals.

Scientific writing for graduate students: a manual on the teaching of scientific writing (by F.P. Woodford, 1968; Rockefeller University Press, New York; 190 p.) is intended primarily for the *teacher*, and for writing and editing in the life sciences. However, earth scientists may profit from it. Most of the text concerns the writing of a journal article; other chapters cover topics like design of tables and figures, oral presentation of a paper, and how to search the literature.

Effective is the word for *Effective writing for engineers, managers, and scientists* (H.J. Tichy, 1967; John Wiley & Sons, New York; 337 p.). If you must settle for a single book on the subject, this one is it. One chapter alone will convince almost anyone: "Two dozen ways to begin".

Another good one, closer to the earth sciences, is *Engineered report writing* (by Melba W. Murray, 1969; Petroleum Publishing Co., Tulsa, Okla.; 121 p.). A sample of the author's message may be found in *Geowriting*'s section "A style of your own".

Outside the earth sciences but worthwhile nevertheless is *Scientific writing* (by Lester S. King and Charles G. Roland, 1968; American Medical Association, Chicago; 132 p.). This is an outgrowth of a series of short courses and workshops covering many phases of medical writing and editing. Apparently the faults and problems of medical writing are much the same as those of geological writing, and so geologists may learn much from these chapters on such subjects as monotony, the passive voice, "Why Not 'I' and 'We'?", jargon, and "Making It Shorter—or Longer".

copywriting Another non-geology subject is covered in *The compleat copywriter* (by Hanley Norins, 1966; McGraw-Hill, New York; 326 p.), "A comprehensive guide to all phases of advertising communication". Even if you never have anything to do with advertising, Norins will help you reach your reader.

To many editors (and writers, and even those who only read), usage is the most fascinating of subjects. Almost everyone has his own notions on the subject, and many follow the practice of making up their own lists of usages to guard against. As a result there are a great many book-length treatments of the subject and several that are outstanding.

usage—and abusage

The classic work on English usage, and the most widely used reference in editorial offices throughout the English-speaking world, is *A dictionary of modern English usage*, known as "Fowler" or "M.E.U." (H.W. Fowler; revised by Sir Ernest Gowers, 1965; Oxford University Press, Oxford; 725 p.). Few editors dare neglect it, and fewer still will do so after reading entries on (say) "humour" or "split infinitive".

There are those who believe that Fowler's usage is excessively British, and *A dictionary of American-English usage, based on Fowler's Modern English Usage* (by Margaret Nicholson, 1957; Oxford University Press, New York; 671 p.) was intended to correct that supposed defect. In fact, Nicholson lacks Fowler's flair and at least two American works are superior to this substitute. It does have the advantage of being available in a paperback edition (1958; Signet, New York; 671 p.).

The careful writer: a modern guide to English usage (by Theodore M. Bernstein, 1965; Atheneum, New York; 487 p.) is an alphabetically arranged discussion of usages. Its origins may be traced to *Winners and sinners*, "a bulletin of second-guessing" that the author has produced for many years from a corner of the *New York Times* news room, with the *Times* itself under review. This work is less exhaustive than Fowler but more entertaining, and it focuses on everyday problems in current American writing.

Bernstein has produced two other books based on *Winners and sinners: Watch your language* (1958; Channel Press, New York; 276 p.) and *More language that needs watching* (1962; Channel Press; 108 p.). Another work is his *Miss Thistlebottom's hobgoblins* (1971; Farrar, Straus & Giroux, New York; 260 p.), which is subtitled *The careful writer's guide to the taboos, bugbears and outmoded rules of English usage*. It exposes such superstitions as the rule against splitting infinitives, and also includes several short works on usage such as Ambrose Bierce's *Write it right*.

Another American rival of Fowler is *A Dictionary of contemporary American usage* (by Bergen Evans and Cornelia Evans, 1957; Random House, New York; 567 p.), one of the most useful works of its kind. The Evanses often distinguish between American and British usage as well as between various levels of usage.

Wilson Follett's *Modern American usage* (edited by Jacques Barzun and others, 1966; Hill & Wang, New York; 436 p.) also aspired to compete with Fowler but has proved to be more prescriptive and conservative—some say reactionary.

Another is *A dictionary of usage and style* (by Roy H. Copperud, 1964; Hawthorne Books, New York; 452 p.). Copperud remarks in his preface that "The best a dictionary like this can do is to develop the user's critical faculty"—an admirable goal.

If you're confused by that list you might try another book by Copperud, *American usage: the consensus* (1970; Van Nostrand Co., New York; 292 p.), in which he lists alphabetically many controversial points of usage and compares the opinions of Fowler, Bernstein, the Evanses, Follett, and himself.

literary style

Turning now to literary style: the first book on almost anyone's list is *The elements of style* (by William Strunk Jr and E.B. White, second edition, 1972;

Macmillan, New York; 78 p.). The sections on grammar, style, and usage may seem overly conservative to some, but White's chapter on writing is among the most readable in the language.

"Strunk and White" is a revival, by White, of a textbook in the English department at Cornell dating back to 1919 or earlier. A similar revival, less noted, is *The golden book on writing* (by David Lambeth and others, 1963; Viking Press, New York; 81 p.—originally published by Dartmouth College in 1923). This an old classic, now rather dated, has sound advice on points seldom discussed elsewhere—a fine supplement to Strunk and White.

Careful writers tend to despise what Americans call bureaucratese or gobbledygook, and anyone in that group will be interested in *The complete plain words* (by Sir Ernest Gowers, 1954; Penguin Books, Baltimore; 272 p.). It was written at the invitation of the British Treasury, and is far livelier than most commissioned works. That guide concerns "the choice and arrangement of words in such a way as to get an idea as exactly as possible out of one mind into another."

The five clocks (by Martin Joos, 1967; Harcourt, Brace, New York; 108 p.) is "A linguistic excursion into the five styles of English usage" and one of the most perceptive of all discussions of literary style. Its analysis discloses a structure much more complex and interesting than the conventional ladder-like model of the vulgar, colloquial, standard, and formal. That may sound pedantic; the book is not.

Writers and editors who concern themselves with readability levels should at least know about *The art of readable writing* (by Rudolf Flesch, 1949; Harper, New York; 237 p.). Flesch attempts to measure readability and human-interest levels, and even if you don't want to go to the trouble to make the tests you will find much good advice on how to write for the reader.

Roget's international thesaurus (P.M. Roget, third edition, 1962; Thomas Y. Crowell Co., New York; 1,258 p.) stands in one form or another on almost everyone's reference shelf. In fact it is so widely known that a warning is in order: despite widespread belief, this is *not* a dictionary of synonyms, and *must* be used in conjunction with a good dictionary. The proper and intended use of Roget is to guide the user from a concept (he cannot think of the word) to the word itself.

Mind the stop (by G.V. Carey, second edition, 1971; Penguin Books, Baltimore; 126 p.) is a sensible and readable guide to punctuation and one of the very few of its kind.

geological specialties

John L. Ridgway's *Preparation of illustrations for reports of the United States Geological Survey, with brief descriptions of processes of reproduction* (1920; U.S. Government Printing Office, Washington, D.C.; 101 p.) is badly out of date but has much valuable information that is not easily available elsewhere. The U.S. Geological Survey currently uses unpublished in-house information on the same subject; Ridgway's book has not yet been supplanted.

A specialized geologic topic is treated in "Suggestions for preparation of regional stratigraphic cross-sections" (by J.C. Maher and L.H. Lukert, 1955; American Association of Petroleum Geologists *Bulletin*, v. 39, p. 1,655–1,667).

There seems to be no book other than Ridgway's dealing with geological illustrations only, but Frances W. Zweifel's *A handbook of biological illus-*

tration (1961; University of Chicago Press) may help earth scientists, too.

Several style manuals, such as the ones issued by Baylor University's Department of Geology, and the Geological Survey of Canada, carry information on the preparation of illustrations. Charles Collinson's article, "Size of lettering for text figures", published in the *Journal of paleontology* (v. 36, p. 1,402) may help with that facet.

photos Photographs pose another problem. Two books that deal with technical aspects of photography are Alfred Blaker's *Photography for scientific publication* (1965; W.H. Freeman & Co., San Francisco; 158 p.) and *Photography for the scientist*, edited by C.E. Engel (1968; Academic Press, New York; 606 p.). A vast amount of specialized information is published by Eastman Kodak Company on all phases of photography including the making and presentation of slides for technical talks. Write to Kodak (343 State Street, Rochester, N.Y., 14650) for the current *Index to technical publications*, which lists specialized booklets that may help you. Kodak also has several technical booklets on making slides for lectures and exhibits. For the same purpose, *Slide manual* is published by the American Association *of Petroleum Geologists* (1970; Tulsa, Okla.; 32 p.); careful use of it or one like it should win the gratitude of audiences everywhere—especially those who have been subjected to illegible and unintelligible slides.

Editors and authors alike should be aware of the legal aspects of photography. *Photography and the law* (by George Chernoff and Hershel Sabin, fourth edition, 1971; Amphoto, Garden City, N.Y.; 158 p.) deals with the ownership of photographs; loss and damage to film; law of copyright and libel; and with the distinction between the right to take a photograph and the right to publish it.

editing Even editors sometimes wonder how other editors work. At least 25 answers may be found in *Editors on editing* (edited by Gerald Gross, 1962; Grosset & Dunlap, New York; 265 p.) as there are that number of chapters by that number of editors, discussing such topics as copy editing, textbook editing, and dealing with authors, as well as the famous "Theory and practice of editing *New Yorker* articles" by Wolcott Gibbs. This book reveals many points of view about a single craft.

Much of editing consists of design and production. A concise but wide-ranging introduction may be found in *Typographics: a designer's handbook of printing techniques* (by Michael Hutchins, 1969; Studio Vista/Reinhold, New York; 96 p.). It covers production from communicating with the printer through various problems of binding (and their solutions).

Another very useful small book is *Pocket pal: a graphic arts digest for printers and advertising managers* (1973; International Paper Co., New York; 180 p.). The main title is unfortunate but descriptive: you really can put it in your pocket. A somewhat similar book is *Graphics handbook* (by Ken Garland, 1965; Reinhold/Studio Vista, New York; 96 p.); it "attempts to define the media which are at the disposal of the graphic designer, to clarify the nature and scope of visual information as it relates to his work, and to supply in tabular form hard facts which are essential to any serious effort in this field."

Ink on paper (by Edmund Arnold, 1963; Harper & Row, New York; 323 p.), subtitled *A handbook of the graphic arts*, gives more detail but is not so handy. It tells little about ink or paper, oddly enough, but a great deal about how to get the former on the latter.

Of the many introductions to typography (many good ones may be found in books of larger scope, such as *Ink on paper* and *Bookmaking*), one of the best is *Typography: basic principles* (by John Lewis, 1964; Reinhold, New York; 96 p.). It is subtitled *Influences and trends since the 19th century*, and has useful sections on mechanics and design.

design An editor needs to maintain consistency of design throughout a single publication, or throughout several publications, or even throughout a larger scheme of things. Principles, problems, and solutions are discussed in *Corporate design programs* (by Olle Eksell, 1967; Reinhold/Studio Vista, New York; 96 p.) in terms of coördinating graphic design throughout all aspects of an organization's image—trademark, logotype, typography, letterhead, packaging, and the like.

A more specific problem is attacked in *Bookmaking: the illustrated guide to design and production* (by Marshall Lee, 1965; R.R. Bowker Co., New York; 309 p.). Lee specifies that his subject, which encompasses design and production, has one function: "to transmit the author's message to the reader in the best possible way." This superb work is useful to editors of magazines and other publications as well.

Magazine design (by Ruari McLean, 1969; Oxford University Press, London; 354 p.) shows scores of examples of covers, contents pages, article layouts, and so on, illustrating the problems magazine designers have faced and how they solved them.

For a textbook for an elementary design course, and a valuable aid to breaking out of conventional concepts, try *Principles of two-dimensional design* (by Wucius Wong, 1972; Van Nostrand Reinhold Co., New York; 77 p.).

copyright; law In theory, editors of scientific works should have fewer legal problems than editors of, say, crusading newspapers. However, copyright alone is enough to justify another book or two on the reference shelf. A standard here is *A manual of copyright practice: for writers, publishers and agents* (by Margaret A. Nicholson, 1956; second edition, Oxford University Press, New York; 273 p.).

Another widely used text is *Say it safely: legal limits in publishing, radio, and television* (by Paul P. Ashley, 1969; fourth edition, University of Washington Press, Seattle; 181 p.). This one is primarily concerned with the everyday publishing problems in the fields of libel, contempt of court, and right of privacy.

management A special field for editors is covered by a monthly journal: *Folio: the magazine for magazine management* (Market Publications Inc., New Canaan, Conn.). This seems to be the only one of its kind; it deals not only with management but also with editing processes, circulation procedures, production methods, Postal Service regulations, and on and on.

advertising If you have anything at all to do with advertising, you should see *Advertising age* (published weekly in Chicago); it is all but indispensable for keeping up with trends in marketing, advertising practice, and the like.

dictionaries The essential reference for any writer or editor is a dictionary. A good one can even substitute (up to a point) as a general style manual by serving as a standard for spelling and hyphenation, tabulating proofreaders' marks, discussing principles of usage and punctuation, and defining at least the basic printing terms.

Fortunately there is a wide range of dictionaries. It is too easy to judge them solely by the copyright date on the assumption that the most recent will have the latest definitions and the newest words. Other factors may be quite as important: the presence—or lack—of place names, systems of measurement, and typographic conventions. Also the philosophy of the editors and their basis for choosing terms and defining them is important; the introduction to a dictionary usually presents a forbidding appearance but reading it will be well worthwhile.

In many ways the leading English dictionary is the *Oxford English dictionary*, or O.E.D. (edited by James A.H. Murray and others, 1961; Oxford University Press; 13 volumes, 16,569 p.). The O.E.D. is intended to give each meaning of every word that was ever in English, and to quote the earliest known use of each. The whole is now available in a small-print edition (2 volumes, 4,134 p.); there is also a series of supplements to bring the work up to date. Also there is a long series of Oxford dictionaries (*The Oxford universal dictionary on historical principles*, *The shorter Oxford dictionary*, *The concise Oxford dictionary*, and so on) working down from the 13 volumes plus supplements to one of pocket size.

Perhaps the leading "unabridged" dictionary in the United States is *Webster's third new international dictionary* (edited by Philip B. Gove, 1964; G.&C. Merriam Co., Springfield, Mass.; 2,662 p.). This one is often compared with Webster 2, or the second edition of *Webster's new international dictionary of the English language* (edited by W.A. Neilson and others, 1957; 3,194 p.); many users favor Webster 2 over Webster 3, largely because the second edition specifies "proper" usage and the third is content to report usage. Webster 2 does have several advantages, but it is out of date, out of print, and generally unavailable. Both the O.E.D. and Webster give definitions for a given term in chronological order; that is, the last definition is the most recent and therefore often the most preferred.

Now, desk dictionaries: *Webster's new collegiate dictionary* (edited by Henry Bosley Woolf and others, 1973; Merriam; 1,536 p.), the eighth *Collegiate*, is based on the 10 million cards used in compiling Webster 3 and 1 million cards gathered *since* Webster 3. Thus it is more up to date than its ostensible parent, and, in addition, uses different (and generally thought superior) rules for hyphenation.

At least two dictionaries have been published partly as a result of the controversy about Webster 3's "permissiveness". *The Random House dictionary of the English language—the unabridged edition* (edited by Jess Stein and others, 1968; Random House, New York; 2,059 p.) and its *College edition* (otherwise the same title, edited by Laurence Urdang and others, 1968; 1,568 p.) specifies whether a word or usage is "standard". A different approach is taken by *The American Heritage dictionary of the English language* (edited by William Morris, 1973; Houghton Mifflin Co., Boston; 1,550 p.), which was compiled with the aid of about 100 persons constituting a "usage panel", resulting in such rulings as this one, under the entry "anxious"; "The example *anxious to see your new car* is unacceptable in writing to 72 per cent of the Usage Panel, but acceptable in speech to 63 per cent."

Other recent desk dictionaries worth considering include *Funk & Wagnalls standard college dictionary* (edited by Ramona R. Michaelis and others, 1963; Funk & Wagnalls Co., New York; 1,606 p.), which stresses its usage notes; and *Webster's new world dictionary of the American language* (edited by Joseph H. Friend, David B. Guralnik and others; second college edition, 1970; World Publishing Co., Cleveland; 1,692 p.), whose editors stress new science terms and etymologies—both valuable points to consider.

For background on the Webster 3 dispute, see *Essays on language and usage* (edited by Leonard F. Dean and Kenneth G. Wilson, 1963; second edition, Oxford University Press, New York; 409 p.), which includes sections on dictionaries, usage, style, and history and structure of English. It is valuable for long discussions of Webster 3, its merits (by Bergen Evans) and demerits (Dwight MacDonald).

Any dictionary-user will find interesting and useful *The gentle art of lexicography as pursued and experienced by an addict* (by Eric Partridge, 1963; Macmillan, New York; 119 p.), especially chapters on alphabetical order and the problems of assemblage.

glossaries　　A writer or editor in earth science is likely to need dictionaries on special subjects, often called glossaries. Perhaps the most general of these special dictionaries is the *Glossary of geology* (edited by Margaret Gary, Robert McAfee Jr., and Carol L. Wolf, 1972; American Geological Institute, Washington, D.C.; 857 p.), with about 33,000 terms in geology and related earth sciences. Definitions are descriptive, not prescriptive, but in cases of so-called misuses such as "tidal wave" (for "seismic sea wave") the preferred term is also cited. A 52-page bibliography lists many specialty dictionaries and other reference works.

A dictionary of mining, mineral, and related terms (edited by Paul W. Thrush and others, 1968; U.S. Bureau of Mines, Washington, D.C.; 1,269 p.), with about 55,000 terms, might seem to be much like the *Glossary of geology*; however, it "defines" terms only by citing quotations, and it omits petroleum terms and embraces non-geologic fields such as metallurgy and ceramics.

scientific names　　New mineral names are approved by the New Minerals & Names Committee of the International Mineralogical Association. The committee also passes on the validity of mineral species. Published summaries of their findings are included in the journals *American mineralogist* and the *Mineralogical magazine*. Also, see *An index of mineral species and varieties* arranged chemically (by M.H. Hey, 1955; second edition, British Museum [Natural History], Department of Mineralogy, London; 728 p.) and its appendix (1963; 135 p.). Also: *Glossary of mineral species, 1971* (by Michael Fleischer, 1971; Mineralogical Record Inc., Bowie, Md.; 103 p.), an alphabetical list of mineral species as of July 1, 1971, with chemical formulas. The *Mineralogical record* magazine carries lists of new minerals and discredited names.

In American stratigraphy, an indispensible reference is *Lexicon of geologic names of the United States* (edited by M. Grace Wilmarth, 1937; Bulletin 896, U.S. Geological Survey, Washington, D.C.; 2 volumes, 2,396 p.). The *Lexicon* consists of rock-unit names recognized by the Survey, together with age, type section, source of original description, and similar data. (Although not intended as such, it is a useful reference for spellings of geographic names). "Wilmarth" has been supplemented but not supplanted by *Lexicon of geologic names of the United States for 1936–1960* (edited by Grace C. Keroher, 1966; Bulletin 1200, U.S. Geological Survey; 3 volumes, 4,341 p.) and at least one later volume (Bulletin 1350; 848 p.).

Rules for naming a rock unit are strict, and geologists should comply with them. The general rules are given in the *Code of stratigraphic nomenclature*, published by the American Association of Petroleum Geologists (1970; Tulsa; 22 p.).

There is no "rule board" sitting to pass on the use of formation names. Each geologist (unless he works for the U.S. Geological Survey or is publishing with another stringent organization) is on his honor to follow the stratigraphic rules if order is to prevail. Editors must check to be sure that names are used properly, and that new formation names follow the guidelines laid down by the Commission.

The rules for zoology and botany are older, and reasonably clearly defined. There are separate codes for zoology and botany, as well as special codes within these general fields, including special codes for horticulture, bacteriology, ornithology, entomology, arachnology, and a now-obsolete code for paleontology. In order to name a fossil or verify a fossil name, one should refer not to this two-page code, written in 1881, but to the paleontological section of the botanical or zoological codes. A particularly useful book to guide you through the maze is *Naming the living world* (by Theodore Savory, 1962; English Universities Press Ltd., 102 Newgate St., London; 128 p.). Savory's book tells you how to find rules and names; it does not give details of the rules, or lists of species names.

geography Place names of most large landscape features are to be found on the topographic maps of the U.S. Geological Survey. In addition, there are, of course, names given to streets, buildings, and the like within incorporated or populated areas, although these rarely concern the geoscientist. If they do, city or county records are the most accurate source of names.

When in doubt about the correctness of a name, or if you want to create a place name, the United States Board of Geographic Names is the final resort. Decisions of the Board are published quarterly under the title *Decisions on geographic names in the United States.* In it, approved new names, changes in names, and vacated names are given alphabetically by state. Information about the Board's work, or about geographic names, may be obtained from the U.S. Geological Survey (Reston, Va., 22092), whose staff provides assistance to the Board.

Among other special-interest works are the *International dictionary of geophysics* (edited by S.K. Runcorn, 1967; Pergamon Press, Oxford; 2 volumes, 1,728 p.), and *Glossary of oceanographic terms* (edited by B.B. Baker Jr., and others, 1966; Special Publication 35, U.S. Naval Oceanographic Office, Washington, D.C.; 204 p.).

atlases Good atlases, unlike good dictionaries, do not abound. *The Times atlas of the world* exists in several editions, including the 5-volume Mid-century edition (1960; edited by John Bartholomew) and the single-volume 1967 edition, which is cheaper, handier and more complete. The most recent edition (1973; Quadrangle Books, New York; 267 p.) is by both the *Times of London* and the *New York Times*; it is the best general English-language atlas available today, regardless of price. The *National Geographic atlas of the world* (edited by Franc Shor, 1960; National Geographic Society, Washington, D.C.; 343 p.) is widely used and accurate. Geographic names are covered by *Webster's new geographical dictionary* (1972; G.&C. Merriam Co., Springfield, Mass.; 1,370 p.). An outstanding but expensive atlas is the *National atlas of the United States of America* (edited by Arch C. Gerlach; 1970; U.S. Geological Survey; 417 p.).

symbols Symbols and abbreviations help geologists communicate. A recent trend is to discourage writers from using abbreviations. There are notable exceptions: for example, the use of DNA and RNA in place of the long names for those acids makes it easier to read articles where those terms are often repeated. Abbreviations and other symbols are commonly used to denote units of measurement (linear, time, volumetric, and others); chemical and biochemical elements, compounds, and components; and maps.

One of the handiest sources of symbols and abbreviations in science is the *Handbook of chemistry and physics* (latest edition: the 53d, 1972; Chemical Rubber Company, Cleveland; about 2,200 p.). Among many other tables are mathematical, chemical, and physical signs, symbols, and abbreviations; Greek and Russian alphabets; American standard abbreviations for scientific and engineering terms, and abbreviations of common units of weight and measure, recommended by the National Bureau of Standards (in Miscellaneous Publication 286; 1967).

Many of these tables are also available in desk and unabridged dictionaries.

Other sources of symbols and abbreviations, many of whose products are included in the *Handbook*, are the International Organization for Standardization, Geneva, Switzerland, and the American National Standards Institute (1430 Broadway, New York, 10018).

Abbreviations and symbols for the International System of Units (SI) are defined in "Metrication in scientific journals" by the Royal Society Conference of Editors (1968; *American scientist*, v. 56, p. 159–164). Those units should now be used wherever possible (however, a common error is to convert units with excessive—false—precision). SI units are basically permutations, combinations, and decimal multiples (metric units) of the defined standards for length (meter), time (second), mass (kilogram), thermodynamic temperature (degrees Kelvin), electric current (ampere), and luminous intensity (candela). There is only one SI unit for any quantity. Multiplication or division of SI units produces other SI units, some with special names; hence force is measured in newtons, energy in joules, power in watts, pressure in pascals, magnetic flux in webers, and magnetic-flux density in teslas.

For convenience, multiples of these units may be expressed as powers of 10 using the prefixes given here (as applied to linear distance).

Many "Where to find information on" books and booklets have been published in geoscience. They are very useful, but as the information they refer to is constantly changing, their useful life is very short. One of the most recent is *Geologic reference sources: a subject and regional bibliography of publications in the geological sciences* (by Dederick C. Ward and Marjorie W. Wheeler, with Mark W. Pangborn Jr., 1972; Scarecrow Press, Metuchen, N.J.; 453 p.). It provides an introduction to the literature of geology and to publications and maps dealing with the general geology of a country or region.

A particularly useful "where to" book is *A bibliography of earth science bibliographies* (compiled by Harriet K. Long, 1971; American Geological Institute, Washington, D.C.; 19 p.).

For information that ranges from physical science and engineering, through social sciences, the federal government, water, and toxicology, the Library of Congress National Referral Center has published specialized directories. These, like *A directory of information resources in the United States: biological sciences*, are available from the U.S. Government Printing Office.

To check quotations, first try John Bartlett's classic *Familiar quotations: a*

distance	meters	abbreviation
geometer	10^{20}	Gmm
terameter	10^{12}	Tm
gigameter	10^{9}	Gm
megameter	10^{6}	Mm
myriameter	10^{4}	Mym
kilometer	10^{3}	km
hectometer	10^{2}	hm
decameter	10^{1}	dam
meter	10^{0}	m
decimeter	10^{-1}	dm
centimeter	10^{-2}	cm
millimeter	10^{-3}	mm
micrometer	10^{-6}	μm
nanometer	10^{-9}	nm
Ångström	10^{-10}	Å
picometer	10^{-12}	pm
fermi	10^{-13}	F
femtometer	10^{-15}	fm
attometer	10^{-18}	am

"Where to?"

collection of passages, phrases, and proverbs traced to their sources in ancient and modern literature 14th edition, (1968; Little, Brown & Co., Boston; 1,750 p.). Another help in tracking down an elusive quotation is the *Oxford dictionary of quotations* (1955; second edition, Oxford University Press, Oxford; 1,003 p.).

Other helps include almanacs (several such as the *World almanac* are published annually), encyclopedias, concordances, indexes, directories, thesauruses, guides to archival and unpublished material, lists of theses and dissertations, lists of publications from various publishers—federal, state, society, and private—specialized glossaries and dictionaries, special atlases, and many, many more, available generally at libraries. Consult your librarian for sticky problems, especially for areas not in your field.

about people
For biographical information about scientists, details are given in *American men and women of science* (12th edition, 1973; Jaques Cattell Press/R.R. Bowker Co., New York). For a few scientists and many others, see the various versions of *Who's who*, including *Who's who in America*, *Who's who in the West*, *Who was who* (published by Marquis Who's Who Inc., Chicago); most are published at 2-year intervals. Certain professions and specialties have special *Who's whos*, such as *Who's who in engineering*, *Electrical who's who*, *Who's who in government*, and *Who's who in politics*.

British biography is included in *Who's who* (St. Martin's Press, New York), published annually, and the *Dictionary of national[British] biography* (Oxford University Press). American biography is not easy; try *Webster's biographical dictionary* (1972; G.&C. Merriam Co.; 1,697 p.), which is frequently revised so that it may be necessary to refer to older editions (which Merriam calls printings) to locate a person who has retired or dropped from the limelight. *Dictionary of American biography* (G.C. Scribner's Sons, New York), issued in several volumes, with supplements, may also help. *Who's who in the world* (A.N. Marquis), *Who's who in science in Europe* (Francis Hodgson Guernsey), and *Who's who in the U.S.S.R.* (Scarecrow Press, New York) include names not listed elsewhere.

Memorials published by the Geological Society of America, the American Association of Petroleum Geologists, and other professional organizations are good sources of information about deceased geologists. The *New York Times* obituaries are useful sources. (The *Times* Information Bank makes all its articles accessible at remote computer consoles.)

indexes
Indexing is an art that requires high skill. Purpose of the index, interests of the users, subject matter, bias of the indexer—all those enter into the making of an index. There are rules one can follow: libraries have special requirements; individual publishers often follow individual methods; society publications and technical journals frequently have in-house requirements that have been developed through the life of the publication.

The U.S. Geological Survey has published the scheme of subject indexes used in preparing the *Bibliography of North American geology* and its abstract journals. Although those journals are no longer published, the general philosophy of indexing earth-science papers is reflected in the description given in the hard-to-find *Guide to indexing bibliographies and abstract journals of the U.S. Geological Survey* (77 p.).

Study of good indexes and practice in making your own are probably the best ways of learning to index. There are books and articles on basic indexing techniques available at libraries, but none exists on the specific problems involved in **indexing books** or other publications in earth science.

For general purposes, try *Indexing: principles, rules and examples* (by Martha Thorne Wheeler, 1957; fifth edition, New York State Library, Albany; 78 p.), which includes a long bibliography on the subject.

Authors and editors may find their way through the maze of making a normal index, but they are unlikely to be able to do key-word indexing for storage in a computer unless they have some reasonably specific information on what is required in making and using a computer-stored index.

The design and use of computer-system indexes involve a whole new set of concepts and techniques. Most machine systems require artificially brief entries—spawning semi-coded index "languages" like "Kwik"—as well as a standard but finite list of selected terms that are established in advance as acceptable for the given index system.

Intended to be a uniform base for indexing the various branches of engineering and science, the *Thesaurus of engineering and scientific terms* has been published by Engineers Joint Council (345 East 47th Street, N.Y., 10017). The book's subtitle describes its intent: *A list of engineering and related scientific terms and their relationships for use as a vocabulary reference in indexing and retrieving technical information.* Using the *Thesaurus of engineering and scientific terms* as the basic list of machine-acceptable reference words, the indexer can construct more detailed breakdowns of specific fields, thereby creating a specialized, computer-oriented thesaurus. The International Mineralogical Association and the U.S. Office of Water Resources Research have done just that: prepared key-word thesauruses for use by the editors and indexers of all journals dealing with their subjects. By the time you read this, thesauruses in other fields may be available.

looking ahead

Books are here to stay, but
* *microforms are becoming more important.*
The trend is toward
* *universal indexing systems and*
* *individualized, computer-managed systems for input/output.*
However, good writing, careful editing, and accurate input remain most important.

By the end of this century, it seems likely that writers and editors will be dealing with ● microforms as primary bibliographic tools and documents, ● documents that are largely converted to machine-readable forms, ● universal standards for codification and indexing, ● automatic abstracting and titling, ● machine-produced citations, and ● automatically produced documents.

That glimpse suggests the kind of change that is upon us. Even at this writing a manuscript can be stored on tape; an editor can recall any paragraph for projection on a television-like screen; using a typewriter-like keyboard, he can select any given line on the screen and correct, rewrite, or delete it from the tape. He can add type specifications to the tape, and within hours complete camera-ready pages of an entire book can be ready for plate-making. Also, presswork and binding have been automated so much that paper goes in one end of a book-making machine and bound copies come out the other.

However, some things cannot be automated. Good writing and sound editing are more important than ever. If they are not done right—before input—much of the speed and economy of technological advances are wasted.

From an information point of view, too, we are entering an entirely new information-transfer environment. Information systems such as the computer-based bibliographic services are moving rapidly toward custom designing for each user or each specific problem.

Example: one science publisher, who sells separates of individual papers, uses the record of each sale as feedback to refine an extensive profile of customers. The profile makes it possible to determine closely how many copies of a given paper will sell, enabling the publisher to make accurate print orders and reduce postage costs, storage space, and so on.

"Made to order" textbooks are now being assembled from material that specialty publishers have rights to—plus, perhaps, some new material —are now being printed with press runs as small as a few hundred.

Because technological advances reduce response time, the traditional print methods may lose ground to other forms. Many people believe that microforms (especially microfiche) lack only a good portable reader to displace much print. And the combination of condensed size and the new retrieval methods promises savings in time and money as well as enhancing capability.

These new developments add to the stress on preprocessing of input. As such, they might even lead to better "writing".

However, if in 30 to 50 years all professional workers will have at their elbows an on-line access to a universal data base, how much of this book will be worth preserving?

That day seems too far away to support sloppiness in writing while we await the advent of Big Brother. Remember the acronym used by computer specialists: GIGO—Garbage In, Garbage Out.

indexed check list

Publications vary so much that no check list will suit them all. We have itemized many considerations that might have to be made in the course of writing, editing and rewriting, illustrating, designing, copyfitting, and proofing, and producing your work. The numbers shown refer to pages in *Geowriting* where you can get more information about a particular item. If not discussed specifically in *Geowriting,* the reference is to a cited reference or example used.

30